THE STORY OF CHEMISTRY

THE STORY OF CHEMISTRY

From the periodic table to nanotechnology

Anne Rooney

This edition published in 2017 by Arcturus Publishing Limited
26/27 Bickels Yard, 151–153 Bermondsey Street,
London SE1 3HA

ISBN: 978-1-78428-545-6
AD005418UK

Printed in China

Contents

THE STUFF OF THE WORLD

'Chemistry is the study of the material metamorphosis of materials.'

August Kekulé, 1861

Chemistry and magic appear to have a lot in common. Both involve the transformation of matter by invisible means. While chemistry can't change a prince into a frog, it can explain how the matter of food, air and water can be transformed into either a frog or a prince. With chemistry, you can produce substances that make food toxic or taste better and grow crystals seemingly from nothing. Chemistry deals in fires with coloured flames, liquids that change colour and metals that slither like liquid or catch fire. No wonder it has captivated people's imaginations for millennia.

The surprisingly explosive reaction of sodium with water is not how most people would expect a metal to behave in contact with Earth's most common liquid.

The vibrant blue colouring of medieval Islamic tiles is produced by cobalt oxide.

Chemistry is the study of the stuff – matter – that makes up the physical universe; it attempts to explain how and why matter interacts and changes as it does. The story of chemistry begins long before people understood much about the nature of matter, but daily exploited its properties. Our ancestors collected and used their knowledge of what would become chemistry without fitting it into any explanatory, theoretical framework. They found that certain stuff from the earth coloured their glazes blue, or that treating the iron they smelted in certain ways made it stronger. But this was just how things were, beyond rational explanation. Chemical knowledge accrued in the traditions of craftworkers and was passed on for its usefulness.

Elements and particles

The beginning of science, including chemistry, is usually located in the culture of Ancient Greece, more than 2,500 years ago. It was there that people started to seek explanations that were not rooted in the supernatural. As proto-chemists, they

The activity of alchemists centred on trying to turn base metals into gold and silver. As this painting by Pietro Longhi shows, health and safety provision was lax in 1757.

began to account for the behaviour of the material world by recourse to philosophically constructed ideas about the nature of matter. These included the first suggestions that matter might be composed of elements and broken into tiny particles – though their versions of these ideas were far from our own, and competed with other models.

The Ancient Greeks left the world some ideas about elements and particles and an early approach to a scientific way of thinking, but it would be more than 2,000 years before solid progress towards modern chemistry began. In the interim, chemistry was alchemy: the semi-mystical search for agents of transformation that could turn base metals into gold, or bestow health and even immortality. For all its outward appearance, alchemy was not magic. It was founded in solid knowledge of chemicals but interpreted within a flawed framework; false notions followed logically and inevitably.

Chymistry

Even when 'real' chemistry began to emerge during the Scientific Revolution of the 17th and 18th centuries, many rational scientists continued their alchemical research, seeing no conflict between this and their more mainstream investigations. This mix of chemistry and alchemy in the early modern age is often referred to as chymistry. When the current models of atoms, elements and molecular bonds emerged, alchemy and chemistry finally parted company. The central premise of alchemy was no longer tenable. In the spirit of science, alchemy had to be let go.

The story of chemistry is one of a yearning to understand and master the stuff of the world around us. It's a story with a false start, but it still made advances in the details while missing the overall picture. The alchemists and chymists made huge progress in finding out how materials behave and how to make new and useful chemical compounds, and in developing techniques and equipment which are still used today – even though their theoretical framework was badly wrong.

A central role

From the 18th century onwards, with chemists finally on the right track, progress accelerated. Chemistry came of age with the modern paradigm of atoms of the elements combining by forming chemical bonds. At that point, the links between chemistry, biology and physics became clear. Chemistry now occupies a central place in the larger story of science, holding the other disciplines together. Chemists have unpicked the mysteries of matter and can now explain and predict the changes wrought by heating, combining, refining and otherwise messing around with the chemical stuff of the world. The processes that puzzled our predecessors have now largely been explained.

Modern chemistry is still about the transformations of matter, but is rooted in understanding. It works in tandem with and in the service of a host of other disciplines. Chemistry reveals to us the workings of the natural world – including our own bodies – and gives us the tools to make new materials, tailored to our needs, that don't occur in nature. It also gives us the means to wreak terrible havoc. It is our responsibility to use it wisely.

CHAPTER 1

CHEMISTRY WITHOUT KNOWING IT

'In science it is a service of the highest merit to seek out those fragmentary truths attained by the ancients, and to develop them further.'

Johann Wolfgang von Goethe, 1749–1832

One of our defining features as humans is our use of the material we find in the natural world. Since prehistoric times, we have made pigments, tools, foodstuffs, pottery, bricks, medicines, perfumes and jewellery, moulding the matter around us into new physical and chemical forms. We did it long before we had any concept of 'chemistry' as a science.

Making medicines has been an important spur to progress in chemistry for millennia. Here, 13th-century Persian pharmacists are at work.

Adventures in pre-chemistry

Our earliest ancestors began their exploration of pre-chemistry when they discovered the transformative power of fire, or ground minerals and plant matter to make pigments and medicines. These first adventures were doubtless haphazard and random; they would have revealed some substances that were useful, some that were not, and probably some that were downright dangerous.

Early humans messing around with matter and its properties exploited the riches of the natural environment in a way that involved changing them to uncover new properties. This is the very essence of chemistry: to discover the ways in which matter can be transformed and use them to your advantage. We can easily imagine a forebear from the Paleolithic period poking a stick in the fire and finding that he or she could then make a mark on the rock with the blackened end, or that juicy, chewy, fibrous meat becomes easier to eat and has a different flavour after the application of fire. Perhaps dyes and pigments were first discovered accidentally, when squashed plant material left a stain. Without curiosity, these serendipitous accidents would have led nowhere. Inquisitive humans who heated lumps of earth with sparkly seams running through them to extract the metal, or fashioned clay from the mud at their feet into useful shapes, were the first proto-scientists, the first pre-chemists. They didn't know or need to know how the transformation of matter worked or why it changed properties – they simply explored and exploited their discoveries in ways that became essential to human culture and civilization.

Neolithic cave paintings made 2,500–4,000 years ago in Thailand using a bright red pigment.

Chemistry of colours

People began to decorate their environment by painting the walls of caves they lived in a very long time ago. The earliest evidence of pigment-making dates from 70,000–100,000 years ago in the Blombos Cave system in South Africa. Here archaeologists found two ingredients for making paint – ochre and animal bones that artists would simply grind together. Ochre is a naturally occurring mineral consisting of silica and clay and an iron-rich substance called goethite, which gives it its yellow to orange-brown colour. Other prehistoric paints were made from carbon (burnt wood or bones) for black, chalk (calcite, calcium carbonate) for white, and mineral pigments including umber (a natural mixture rich in iron and manganese) for browns and creams. Sometimes mineral pigments might be found in clay that could be applied directly to a surface, like crayons. Otherwise pigments were ground and mixed with water, plant juices, urine, animal fat, egg white or some other carrier that

Ochre from cliffs near Roussillon, France, has been used since prehistoric times; its modern processing to make an indelible dye dates from 1780.

Powdered umber, a mineral pigment from Umbria in Italy.

would evaporate or solidify after the mix had been painted on to the wall. It seems that the earliest reason for mining rock was to extract mineral pigments for painting on cave walls or for body decoration, and people travelled considerable distances to collect them.

Pigments used to dye cloth or to adorn the body were often plant-based. Some of those were not permanent and would wash out in water, so a bit of experimentation would have been needed to discover which were fast and which were not (and 'not' might have been an advantage in the case of body decoration).

From Paleolithic to pots

By the time of the Neolithic period, around 10,000 years ago, people had begun to settle in one place and to farm the land. They soon developed pottery and began working with metals. Both involved processing materials using heat and sometimes mixing them together to change their properties.

Kilns for firing pottery first appeared around 6000BC. Coloured glazes, to give permanent colour to pottery, were first used around the 4th or 3rd millennium BC. They were made by mixing minerals with sand and heating them to melting point. Such glazes may well have been discovered accidentally, as copper smelting was carried out in clay furnaces; a blue glaze could easily have appeared on stones or clay, as copper formed compounds on the surface.

Clay was also used to make bricks, and either left to dry in the sun or baked hard in an oven. The clay was often mixed with straw, which made it stronger.

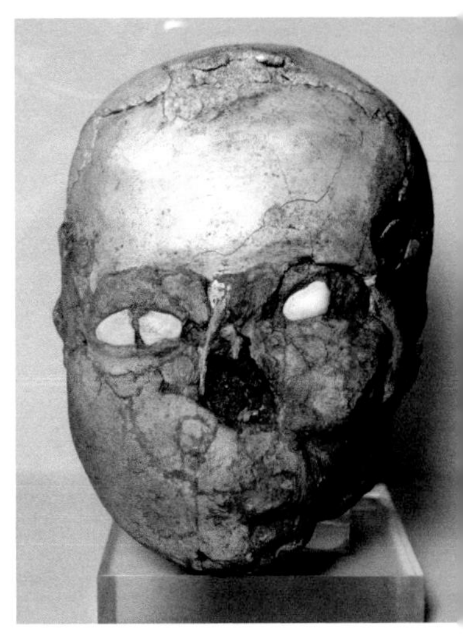

Plaster, pigments and shells (for eyes) were used to recreate the head of a dead person in the earliest known ancestor veneration practices of the Middle East, about 7000BC.

GLASS FROM GLAZES

Glass was first developed as a glaze for pottery around 3500BC in Mesopotamia (present-day Iraq). A thousand years later, it was produced as a substance in its own right. The chief ingredient of glass is silicon dioxide, which is abundantly available as sand. Any civilization with a beach can make glass, and they did – the Ancient Greeks and Romans both made exquisite glassware. The molten glass could be formed around a mould, blown or cast, and was easily mixed with minerals to produce vibrant colours. Glass is not especially robust, but is exceptionally hard and doesn't corrode or dissolve, which made it useful for later forays into chemistry.

The blue-green colour of Egyptian faience is produced by copper pigments.

Metals and mining

The very earliest activity in proto-chemistry left few traces. It takes little effort and no special tools to pick plants and crush them together. Working with clay is easy in areas where clay can be scooped up by hand from riverbeds. The results, painted or dyed artefacts and pottery, don't usually last long. When people learned to fire their pottery in a furnace, rather than drying it in the sun, it lasted longer but was still fragile. Much of the surviving early pottery was deliberately preserved, as grave goods.

Working with metal is more complex, and produces more durable objects. It takes physical work to extract metal and to fashion it, generally requiring high temperatures and some peril. Mines and smelting leave evidence for archaeologists to find.

Six solid metals have been known since prehistory: gold, silver, tin, copper, lead and iron. Copper was used first, and was discovered independently in several places around the world. The first evidence of working with metals is a copper-smelting site in Serbia dating from 5500BC. It took another 1,000 years or so before the discovery that mixing copper with other substances such as arsenic or tin makes a much harder and more useful metal, the alloy bronze. Bronze was soon being widely used for tools and weapons. The manufacture of bronze marks the end of the Stone Age and the start of a new period in human history – the Bronze Age – which began in the 5th millennium BC in the Middle East, India and China.

Bronzes of the Bronze Age

The first bronze was a mix of copper and arsenic. It's likely that arsenical bronze was invented by accident. Copper and other metals generally occur naturally in an ore (rock that contains deposits of a metal or mineral), and sometimes with other elements in a compound. The metal must be separated from the rest of the rock and 'reduced' – that is, have the oxygen

removed. This is achieved by smelting, which involves heating the ore. The metal combines with oxygen from the air, forming a calx (oxide). The calx must then be reduced to free the metal. A common method is (and was) to heat it with charcoal in an oxygen-poor atmosphere. After a fire has burned in a confined space for a while, it will naturally have used up much of the oxygen available, so that would not have taken much experimentation or any chemical knowledge to discover. If the ore contained copper but no arsenic, the process would first produce a copper oxide and then, on reduction, pure copper. But if the ore contained arsenic, as it often did, the arsenic would mix with the copper, producing bronze.

In Ancient Japan, copper was smelted in a depression dug in the ground. Ore was heaped with charcoal and burned, the molten metal falling into the pit. Here, the metalworker protects himself from noxious fumes by covering his face with a cloth.

At 817° C, arsenic has a lower melting point than copper (1085° C) and tends to sublimate (go straight from a solid to a gas). In this case, the gas is toxic. As long as the metallurgist didn't breathe in the emissions and die before the end of the process, at least some of the arsenic vapour would dissolve in the molten copper and produce bronze. This accidental bronze would have been found to be superior (in terms of usefulness) to copper, so the ore that produced it would be used again and perhaps added to other copper-bearing ore to make bronze deliberately.

It is highly unlikely that anyone would have extracted arsenic alone – and lived to repeat the exercise. To isolate arsenic would have required melting it, whereupon it would quickly sublimate and produce poisonous fumes. If these were successfully condensed and trapped, the arsenic would not have been very useful. It's a metalloid, so while it has some properties of a metal it also has non-metal properties. The result of the smelting would have been either a grey powder or a black crystal, depending on how quickly it cooled. These difficulties make it unlikely that early copper-workers added arsenic directly to molten copper to make bronze, but only used arsenic-bearing ores.

Bronze containing tin first appeared around 4500BC in Serbia. Unlike arsenical bronze, the advantage of mixing tin and copper could not have been discovered accidentally as the two metals are not found in the same place. Both metals had to be mined, extracted from the ore and then mixed in a molten state. So, from the beginning, making bronze with tin must have relied on transport and trade networks, involving people from different regions.

DAGGER FROM HEAVEN

When Howard Carter opened the tomb of the Ancient Egyptian King Tutankhamun in 1923, 1,300 years after it had been sealed, many surprises and mysteries confronted the archaeological team. Among them was a small dagger with an iron blade concealed within the king's mummy wrappings. That might not sound unusual, but the Egyptians lived in the Bronze Age, before the time of iron smelting, and there were no local deposits of iron. Furthermore, the iron dagger shows no signs of rust despite its age, a finding explained in 2016 when X-ray fluorescence spectrometry revealed a high nickel content. In fact, the iron for the blade came from outer space. The resourceful and skilful metalworkers had discovered an iron meteorite and hammered a blade from it. The composition of the metal closely resembles that of a meteorite since found in the area around the Black Sea. Dark metal beads of a similar composition have also been found, that are 5,200 years old. The Egyptians called meteoric iron 'metal from heaven'.

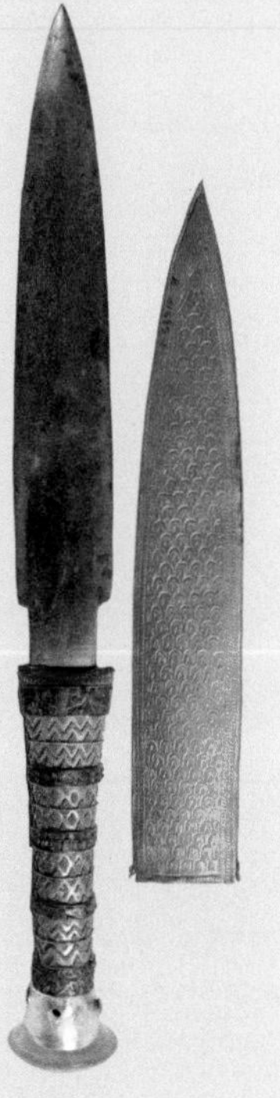

The iron dagger from Tutankhamun's tomb and its decorated gold sheath.

Iron in the Iron Age

Progress in metallurgy was driven by the desire for better farming implements and, particularly, weapons. Later, people learned to extract and smelt iron from its ores. This marked the end of the Bronze Age and the beginning of the Iron Age, generally put at between 1200BC and 600BC in different parts of the world. Malleable meteoric iron was being used in North Africa from at least 3200BC (see box, left).

Iron melts at a much higher temperature than copper, 1538° C, so some major advances in technology were needed before iron smelting could become possible. Although it is called the Iron Age, people did not use pure iron for their tools and weapons. Iron is prone to rust, and is not especially hard. Mixing the iron with the right proportion of carbon makes steel, which is much harder and more resilient. The advent of steel weapons and tools led to rapid advances in technology.

More metals

Gold was probably discovered around the same time as copper, possibly even earlier as it exists in pure lumps in nature. It doesn't have to be smelted from ore, but can be dug from the ground or picked out of rivers as nuggets or dust. It melts at a low temperature and is highly malleable, making it one of the easiest metals to work with. As it doesn't tarnish or react in other ways, it was used for jewellery and other bodily adornments from earliest times; the oldest known gold treasure dates from the 5th millennium BC in Bulgaria. It was even used in dentistry from around 700–600BC by the Etruscans, early inhabitants of Tuscany

The Etruscans made false teeth by fixing human or animal teeth to gold bands, which were then attached to the patient's remaining teeth.

in Italy. Silver was probably discovered soon after copper and gold.

As lead is relatively easy to smelt from ore and soft enough to work, it was used in early times – from at least 6500BC, the date of some cast lead beads found in Turkey. It is too soft to be useful for making tools or weapons and its predominant use in Greek and Roman times was for piping and water tanks. Lead was possibly the first source of anthropogenic pollution; ice core samples from Greenland show elevated lead levels in the atmosphere in the period 500BC–AD300.

Zinc was combined with copper to make brass from the 2nd millennium BC onwards. Zinc-bronze objects from around 1400–1000BC have been found in Palestine, and a prehistoric alloy containing 87 per cent zinc was found in Transylvania. Zinc ores were smelted along with copper, a technique later used by the Romans. Zinc was extracted from zinc carbonate in 13th century India, but had to be rediscovered in Europe, escaping notice until the 1740s.

Harnessing biochemistry

While metal-working was essential to making tools and weapons, many everyday domestic activities made use of chemistry, too. Most of them exploited organic chemistry, which is concerned with carbon compounds, especially those that make up living organisms.

The chemistry of domestic activity – of cooking, dyeing, painting, tanning leather and making medicines, soaps or perfumes – probably developed through simple trial and error. This would soon reveal which pigments are fast and which wash away, which forms of wax or fat make the best lamps and candles, and how the volatile scent of flowers, spices and other substances can be trapped and preserved in waxy or oily mixtures. These processes were used daily and passed on from person to person, often

Brightly coloured pigments are used in modern candle-making.

DEALING WITH THE DEAD

The Ancient Egyptians are most famous for the practice of mummifying their high-status dead. This method of preservation exploited the action of an inorganic compound on the organic compounds of the body.

First, the organs were removed from the body. Then the cavities were stuffed with a naturally occurring substance called natron – a mix of sodium bicarbonate (baking soda), sodium carbonate decahydrate (a kind of soda ash), sodium chloride (salt) and sodium sulphate. (The chemical symbol of sodium, Na, comes originally from 'natron' by way of the Latin 'natrium'.) Natron rapidly dehydrated the body and caused the fats to saponify (turn to soap), preventing putrefaction. The dry, empty body was then stuffed with linen or sawdust mixed with oils and unguents such as myrrh, cinnamon and mastic. Some of these substances had an antibacterial action, helping to prevent decay. The body was wrapped in linen bandages, with oils and resins applied between layers. These made a waterproof seal and again some also had antibacterial effects, further helping to preserve the body.

Natron had many uses outside the funeral parlour. It was used as a cleaning product, to make soap, as toothpaste and mouthwash and as an antiseptic liquid applied to wounds. It was used to bleach cloth, treat leather, as an insecticide and to preserve fish and meat. Natron absorbs water, so it's an excellent drying agent, and in solution produces an alkali which impedes bacterial growth. It was collected from deposits in dry lake beds.

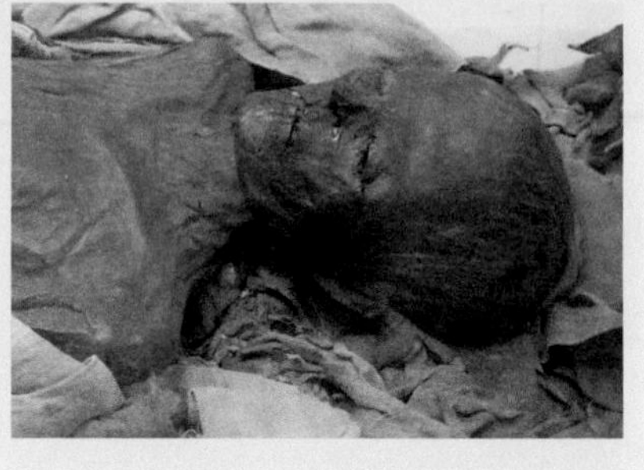

The efficacy of the chemistry of mummification is clear from this unwrapped Egyptian mummy.

The froth on top of brewing beer provided yeast for ancient bakers.

to other, generally illiterate, craftworkers and homemakers. It was not science; it was craft, or even just life.

Making microbes work

Organic chemical processes happen within living organisms. They are the means by which all food is created, whether by plants photosynthesizing or animals processing food, and provide all the functions of living bodies. Two domestic processes make particular use of organic chemistry: brewing and baking both exploit the activity of yeast.

The action of yeast was known at least 5,000 years ago in Ancient Egypt. Naturally occurring microorganisms in or on grains or fruit were probably unseen contaminants that first caused fruit-based liquids to ferment, producing alcohol. There was cross-fertilization between the industries, with bakers taking the froth from beer-brewing to add to their bread, so introducing yeast, which made it rise. Before long, people were taking a portion of fermented drink or dough and using it to seed a new batch, so adding the yeast to kick-start the process (though without being aware of the existence of microorganisms at all). The action of yeast as an organism was not discovered until the 19th century, with the work of French biologist Louis Pasteur (see page 155), by which time these microorganisms had been working for us for several millennia.

The start of chemistry

Our forebears probably thought the properties and reactions of matter such as ores or natron no more remarkable than the changes that occur in meat or eggs when they are cooked. These are also chemical changes, but some that we, too, are so familiar with that we don't think of them as 'science'. At some point, though, scientific thinking began to emerge as people wondered about what happens and why. As with so much else, that point was in Ancient Greece around 2,500 years ago.

Where does it go?

Just boiling a liquid mixture – a pot of soup, say – produces a more concentrated brew by removing some of the water or other solvent. This fact would have been familiar to cooks since prehistory. But when we evaporate a liquid and concentrate or dry out what remains, the liquid must go somewhere. Steam is visible as it evaporates from hot soup, but it then disperses into the atmosphere. The Greek philosopher Aristotle (384–322BC) wrote about the flammable 'exhalation' produced by normal wine. This exhalation is the volatile alcohol, no doubt evaporating quite readily in the hot Greek sun.

It's a small step from noticing that liquid evaporates to finding out how to recondense

This recreation of an ancient Indian method of distillation shows the simplicity of the process. Liquid is heated in a closed vessel and the vapour flows down a pipe to condense in a container, here cooled by running water.

it and collect it. The process of boiling a liquid and collecting the condensate is known in chemistry as distillation. It can be applied usefully to many types of liquid. There is no evidence of the Ancient Greeks using distillation to fortify their wine, despite Aristotle's observation about its 'exhalations'. He did, though, notice that if sea water is heated and condensed, the water collected is drinkable. Alexander of Aphrodisias, in his commentary on Aristotle, mentions that the process was in use in his own time (around AD200). Sea water was boiled in brass kettles and the water condensed on sponges that could then be squeezed out.

Distillation was also used to extract mercury and to produce turpentine. Both procedures were described by Dioscorides and Pliny the Elder in the 1st century AD. Mercury was derived from cinnabar, mined in Spain and used as a red pigment by the Romans. To purify it, they put it into an iron capsule or spoon and heated it in a sealed clay pot. After heating, they opened the pot and scraped the condensed liquid mercury from the inside. They considered this 'artificial' mercury to be superior to that which occurred naturally in the mines, though chemically it was identical. Pine resin was distilled by heating it in a vessel covered with a layer of wool. The condensate, turpentine, collected on the wool and could then be pressed out.

Cinnabar – mercury (II) sulphide – appears as a red deposit in rock.

Doing and thinking

Aristotle didn't only notice the behaviour of evaporating and condensing liquids – he also thought about it at great length. The Ancient Greeks left the earliest record of proto-scientific thinking, beginning with the work of Thales of Miletus (*c.*624–*c.*546BC). He is the first person known to try explaining phenomena by recourse to natural processes rather than the supernatural. People then began to wonder why the materials they worked with behaved as they did. They set out to investigate the nature of matter and its components, at first approaching the problem through philosophy rather than empirical exploration (that is, observation and experimentation). This formed the very origins of chemistry.

Matter and nothing

Before considering the fundamental components of matter, it's necessary to think about whether matter is continuous or divided into particles, though the importance of this might not seem immediately obvious. There were essentially two competing models for the nature of matter: either matter is continuous and fills all of the universe with no gaps, or matter is made of particles punctuating empty space. In Ancient Greek philosophy, this was part of a larger debate – whether everything

is a single unchanging thing, or whether the universe comprises many things and is subject to change.

In the first half of the 5th century BC, the philosopher Parmenides argued that all 'being' is a single, changeless, eternal continuum. The opposite view was developed by Leucippus and Democritus, also in the 5th century BC. They argued that all matter is made of tiny, indivisible particles that exist in a void. They called them 'atomos', or 'uncuttables'. In this model, all matter, and all its properties, is produced by the interaction and arrangement of these atoms. Aristotle rejected this atomic model, as he didn't believe there could be a void; for him, the universe was jam-packed full of continuous matter. His view prevailed until the 18th century, and was used to argue against the existence of a vacuum when Otto von Guericke built a vacuum pump in 1657 (see page 90).

Elementary elements

The next obvious question to ask about matter is – what is it made of?

Again, there were two principal positions in Ancient Greece. It could all be derived from a single original fundamental substance, or it could be made of a mix of a few elements or 'roots'. The older idea was that all matter had a single original form. Thales addressed this, proposing that the source of all matter is water. Anaximenes, another 6th-century philosopher from Miletus, maintained that the origin of everything is air. Observing that water condenses out of air, he proposed that air is the *arche* or original substance from which all others originate. As air cools further, it becomes earth and even stones. At the opposite end of the spectrum, heated air burns to become fire. He was the first to associate the pairs of properties hot/cold and wet/dry as features of matter. From this proposition, it was possible to have a single substance manifest as the various forms of matter simply through displaying different properties.

> *'For it is necessary that there be some nature, either one or more than one, from which become the other things. . . . Thales, the founder of this type of philosophy, says that it is water.'*
>
> Aristotle, *Metaphysics*

Four by four

The Sicilian-Greek philosopher Empedocles (*c.*490–*c.*430BC) formulated instead a scheme based on four elements, which he called 'roots'. It became the basis of Western science for around two millennia. The four roots were fire, air, water and earth, which mingle in different proportions to produce all matter and its various characteristics.

Empedocles countered the view of Parmenides (late 6th to mid-5th century BC) that everything is one and unchanging. He allowed matter to change, and accounted for change by saying that the proportions of the four roots varied. But Empedocles retained one element of Parmenides' model: the four roots themselves could not be destroyed or transformed, they could just change position, becoming separated from one substance and incorporated into another.

The roots were associated with the macrocosm, or the macrostructure of the

ELEMENTS ELSEWHERE

The idea that all kinds of matter are made up of just a few elemental components has been current for millennia and appeared in many civilizations including ancient Babylon, Egypt, China, Japan, India and Greece. Further, most of these civilizations shared most of their fundamental elements. In China, the five elements were earth, water, fire, wood and metal (which was, to all intents and purposes, gold). In India, they were water, fire, air, space and aether or 'void'. The earliest mention of the earth, wind, fire, air grouping is found in Babylon in the 18th to 16th centuries BC, but these elements were not presented as the components of all other matter as they were, later, by the Greeks. In these early accounts, they often had a spiritual aspect or were associated with deities and were not entirely restricted to the physical properties of matter. Even for the Greeks, their elemental nature was invested as much in their properties as in the physical substances, and often a hypothetical pure form was posited, the mundane versions being impure.

The Chinese elements in the sequence of their generation, one from another; from top going clockwise: wood, fire, earth, metal and water.

Earth and universe. At the base or centre is the heaviest component, earth; above it in sequence come water, air and fire. Each element has its own natural place and will tend towards that place, hence fire and hot air rise, water falls through air and earth and stones fall through water.

Aristotle added a fifth element, the rarefied aether found only in the heavens. This was a very fine substance that (he believed) usually existed far above the realms of earth, water, air and even fire. It would come and go through chemistry and physics for 2,300 years – we might not have heard the last of it even now.

Elements and properties

The four roots, which we will start to call elements now, were associated with four properties – hot, cold, dry and wet. Aristotle assigned two properties to each of the four terrestrial elements: fire is hot and dry; air is hot and wet; water is cold and wet; and earth is cold and dry. The fifth element, aether, did not partake of these properties; it was unchanging and pure and didn't enter into combination with any others.

Air and water are fluid and take the shape of any container they are put into; they share wetness. The pairs, fire/water and air/earth, are exact opposites as they share no properties with their partner.

There are two ways of looking at the elements and their properties. One is to see the elements as physical substances that have properties (heat, cold, wetness and dryness) in fixed amounts. Then the proportion of each element in a substance determines its exact nature. For instance, if a substance was made of 80 per cent water

ELEMENTS AND HUMOURS

The human body was generally considered to be a microcosm that mirrored the macrocosm of the universe. It, too, was governed by the same four properties, which were associated with four bodily fluids or 'humours' in a model proposed by Hippocrates (460–370BC). These correspond to the elements according to the properties they share: blood corresponds to air, phlegm to water, yellow bile to fire and black bile to earth. A huge and complex medical model built up around the humours, based largely on the work of the Roman physician Galen (AD130–210). The humoral system endured as the principal model to explain health and disease for around 2,000 years. The system of four corresponding elements and humours created sufficient parallels between the stuff of the world and the stuff of the body to explain, later, the correspondence between models in alchemy and in medicine.

and 20 per cent earth (we might call it mud, but that would be being over-literal), it would have a lot of the property cold, as it gets that from both elements, quite a lot of wet and a bit of dry.

Another way of looking at the elements and properties is less literal. Fire, air, water and earth are not the actual manifestations of the elements we see around us but idealized versions that are carriers for the properties. This makes more sense, as even the Ancient Greeks must have realized they would find it hard to make, say, grapes by roasting earth, water and air together. Yet they could observe that a vine takes something from the heat of the sun, the air and the earth and makes grapes, so it all makes sense if not taken too literally.

Out of Greece

The Greeks, then, established some basic starting points for chemistry. Firstly, four elements make up all the substances found on Earth and account for their properties. There was reasonable consensus about this, and it would be carried forward through Graeco-Roman Egypt, Islamic culture and medieval Europe right up until the mid-17th century. There was less agreement about the second point: whether matter is continuous or made up of discrete particles in a void. Where particles were accepted, it was generally taken that they could not be further reduced than the four elements, so a prototype atomic/elemental framework was in place, at least as one option.

These ideas made chemistry possible. People could begin to explore the stuff of the world and what it can do, giving some kind of rational explanation rather than simply documenting observations or appealing to the supernatural to account for phenomena.

In this medieval depiction of the Tree of Life, Earth is encircled by water, air, and then fire.

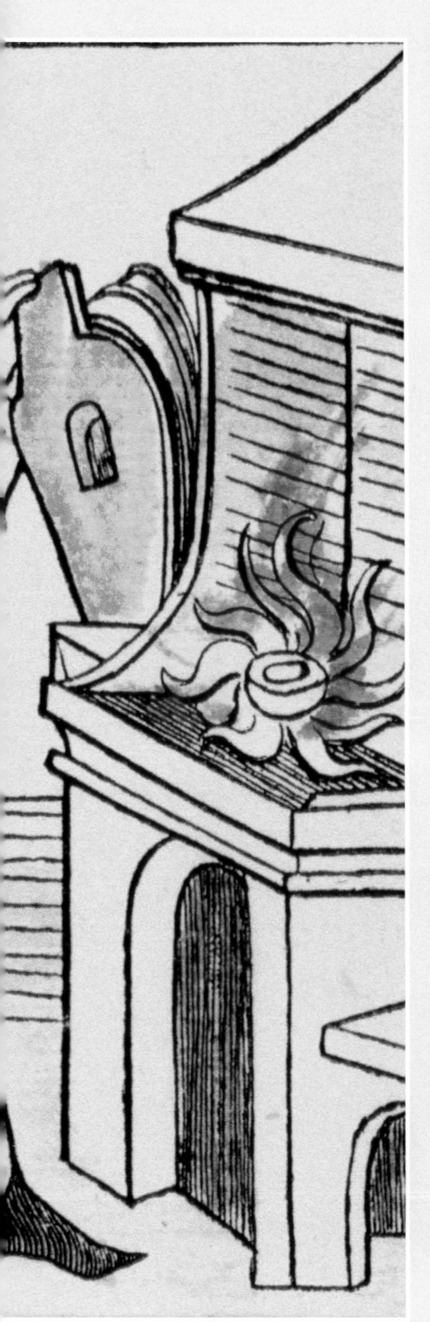

CHAPTER 2

A SPIRITUAL SCIENCE

'Who lets this cursed craft of ours entice him,
The wealth he has in no case will suffice him.
For all the gold or goods that he lays out
Shall be his loss – of that I have no doubt.'

Geoffrey Chaucer, *The Canterbury Tales*, **'The Canon's Yeoman's Tale', 14th century**

The first stirrings of modern chemistry lie in alchemy – a science that was also an art, and which tangled together investigation of the physical stuff of matter with an esoteric spiritual endeavour. To modern eyes, alchemy looks like an attempt at magic. But in focusing on the transformations of matter and the properties that enable transformation, it was very much the first step towards chemistry as we know it today.

An alchemist attempts to turn base metal into gold watched over by a floating (demonic?) figure in this 15th-century woodcut.

A respectable pursuit

No modern scientist could get away with attempting to make magical elixirs or turn lead into gold and at the same time publish respected papers in an academic journal and serve as a Fellow of the Royal Society. Yet that was very much the normal state of affairs in the 17th century. Isaac Newton, one of the most famous scientists of all time, was an alchemist at least as much of the time as he was a physicist and mathematician. There was no conflict between the 'science' of alchemy and any other branch of science. To see why, we need to look back to the origins of alchemy and to examine why and how chemistry eventually broke away from it.

Isaac Newton laid the foundations of modern physics, yet devoted much of his time to alchemy.

The history of alchemy falls into four stages. It has its origins in Graeco-Roman Egypt in the first centuries AD, though few records survive. There is more evidence from the early Islamic period, when alchemy flourished in various Arab lands from the 7th to the 11th centuries. Then it moved to Europe, where the Arab and Greek texts were translated and studied and where, in the 16th century, alchemy took a medical turn. Finally, burgeoning interest in science saw a flurry of activity in the 16th and 17th centuries before alchemy lost ground to modern chemistry.

The shadowy origins of alchemy

It would be rather disappointing if we could pin down the origins of alchemy, isolating a real-world figure and perhaps some down-to-earth manuscript or scroll as the starting point of this otherworldly pursuit. And alchemy does not disappoint in that way. Its legendary origins lie in the 'Emerald Tablet' (*Tabula Smaragdina*) attributed to the Egyptian king Thoth, known to the Greeks as Hermes Trismegistus, or 'thrice-great Hermes'. He is said to have ruled around 1900BC and to have been an extraordinarily wise and skilled individual, well versed in the ways of the natural world. Some legends place him at the time of or even predating the Old Testament figure Moses. He may not have existed at all.

Around a dozen of the works attributed to Hermes Trismegistus have survived and

This imaginative 17th-century depiction of the Emerald Tablet had it conveniently written in Latin.

SNAKE(S) ON A STICK

The principal symbol of alchemy is the caduceus, a rod with two, sometimes winged, snakes twined around it. It was associated originally with Hermes Trismegistus. As the Roman equivalent of the Greek Hermes is Mercury, the caduceus also represents the astrological symbol Mercury (and the planet), and in alchemy represents the metal mercury.

Its use in the USA as a medical symbol is erroneous and results from confusion with the staff of Asclepius (with a single snake, never winged). The caduceus was officially adopted as the symbol of the US Medical Corps in 1902 at the insistence of an officer who wasn't as well informed about classical mythology as he should have been.

The badge of the US Medical Corp shows the caduceus, left, an alchemical symbol.

The Army Medical Department of the US Army uses the more appropriate staff of Asclepius, right.

'Truth! Certainty! That in which there is no doubt!
That which is above is from that which is below, and that which is below is from that which is above, working the miracles of one thing.
As all things were from. Its father is the Sun and its mother the Moon. The Earth carried it in her belly, and the wind nourished it in her belly, as Earth which shall become fire.
Free the Earth from that which is subtle, with the greatest power.
It ascends from the Earth to the heaven and becomes the ruler over that which is above and that which is below.'

'The Emerald Tablet', translated by E. J. Holmyard, 1923

they are a strange jumble, not particularly relating to alchemy. According to legend, the Emerald Tablet was inscribed in Phoenician characters on an actual block of emerald. It was recovered from the dead hands of Hermes (or at least a corpse buried beneath his statue) by Alexander the Great, who ransacked the tomb in the 4th century BC. A legend that places it even earlier has Noah carrying the Emerald Tablet with him on the Ark.

Although the legend is appealing, there is no trace of the Emerald Tablet text before the early 9th century and it seems to be considerably later than the other writings attributed to Hermes (collectively called the *Hermetica*). The cryptic nature of the text provided plenty of scope for later interpreters of an alchemical bent, who

generally took the 'it' referred to as the philosophers' stone, a mythical instrument of transformation (see page 35). There is no good reason to assume that it is about the philosophers' stone, and even if it is, there are many details to be clarified before the text can be considered useful.

A primer in chemistry

The original Emerald Tablet, if it ever existed, is long lost. The first writings in practical chemistry are two papyrus documents dating from the 3rd century AD and are much more mundane, though still fascinating. Written in Greek, they come from Egypt, where the intellectual tradition of Ancient Greece continued under Roman rule, concentrated around Alexandria. They are the only surviving chemical documents from Egypt, the home of the Western alchemical tradition.

The two documents, known as the Leiden Papyrus and the Stockholm Papyrus, contain 250 chemical recipes. These cover techniques that would have been used by craftsmen to make dyes and to make artificial gold, silver and gemstones. They are practical instructions that can still be followed, and share none of the spiritual and mystical paraphernalia of later alchemical writings. There is neither a superstitious nor a theoretical framework for the recipes. For example, one explains how to prepare equal quantities of lime and sulphur with vinegar or urine, heating the mix until it becomes blood-red. This solution, filtered, can be used to colour silver to resemble gold (which it does by depositing a layer of sulphide over the silver). There is no suggestion in the papyrus that this solution actually transforms one metal into another.

First elements and transformations

The four Greek elements of fire, air, earth and water were often represented in a lozenge that shows which properties are shared by each pair of elements (see page 23). Aristotle explained how any element can be transformed into any other by changing the balance of its properties. Such transformations can be effected more rapidly if there are shared qualities:

'The process of conversion will be quick between those which have interchangeable "complementary factors", but slow between those which have none. The reason is that it is easier for a single thing to change than for many. Air, e.g., will result from Fire if a

INTRINSIC AND EXTRINSIC ALCHEMY

Alchemy has often been said to have two overlapping domains. One is extrinsic alchemy, which is the more obviously chemical part and is concerned with the transformation of matter. The other is the intrinsic, or spiritual part. This is largely an artificial modern distinction; for the ancient alchemist, there would have been no division between the two. The methods, aims and practices of the alchemist were rooted in beliefs about the nature of matter and of the relationship between the spiritual, religious and astrological realms and the realm of physical matter. We will try to focus on the practice of alchemy as far as possible and leave the more esoteric mystical aspects to one side, but it is something of an artificial separation.

single quality changes: for Fire, as we saw, is hot and dry while Air is hot and moist, so that there will be Air if the dry be overcome by the moist. Again, Water will result from Air if the hot be overcome by the cold: for Air, as we saw, is hot and moist while Water is cold and moist, so that, if the hot changes, there will be Water. So too, in the same manner, Earth will result from Water and Fire from Earth, since the two elements in both these couples have interchangeable "complementary factors". For Water is moist and cold while Earth is cold and dry – so that, if the moist be overcome, there will be Earth: and again, since Fire is dry and hot while Earth is cold and dry, Fire will result from Earth if the cold passes away.'

Two entirely contrary elements mixed together will produce the opposite pair: '. . . from Fire *plus* Water there will result Earth and Air, and from Air *plus* Earth, Fire and Water.'

Plato, left, and his pupil Aristotle, from Raphael's painting The School of Athens, *1509-11.*

In Aristotle's view, although the elemental properties give matter its form, the fundamental (or primitive) matter was always the same. This is an interesting distinction and one that will recur throughout the history of chemistry. All matter is essentially of one type, but manifests differently, whether through displaying different properties, being differently configured or existing in different conditions. So elemental matter which has the properties dryness and heat, for instance, manifests as the element fire.

This clearly makes transformations between types of matter entirely plausible. If the material itself is not changing but only its properties, it should be possible to change anything into anything else simply by making certain adjustments; that was what the alchemists set out to do. In particular, they aimed to turn cheap metal such as iron or lead into valuable metal – gold or silver.

For Aristotle, the fundamental or 'prime' matter was theoretical. He did not address any possibility of isolating it or working with it directly – all matter around us in the world has 'form' so it has other attributes. But for the later alchemists, prime matter stripped of properties was very much a practical goal and was generally thought to appear as a black lump.

Mixing, making and unmaking

To effect transformations, it seems that all that the alchemist needed to do was to discover the quantities of the elements in,

In a slight misrepresentation of the truth, this 14th-century manuscript shows Aristotle having successfully acquired the philosophers' stone.

say, gold and then match those quantities by meddling with some other matter. It might require reducing the starting material to primitive matter devoid of properties (the black lump) then rebuilding the new matter, gold, by addition or propagation of the needed properties. When the proportions matched, the original matter would have been reformed as gold. What could possibly go wrong? The only tricky thing was – how to do it?

A touch of the vapours

Aristotle believed two 'exhalations' were involved in the production of metals and minerals. One was smoky and the other 'vaporous' and they form when the Sun's rays warm the earth and water. The hot and dry smoky exhalation from warming earth is responsible for the production of minerals, which cannot be melted; the vaporous exhalation from warming water is responsible for the production of metals. Each type of exhalation is impure, being admixed with a little of the other type, so minerals and metals, like everything else, are a mix of the four elements.

He proposed that metals form when the vaporous exhalation is trapped in the earth and put under pressure by the cold, dry element, transforming it. Hence metals are found as seams or ores in the earth. Aristotle believed that metals could grow organically, so a small seed of gold could grow into a larger nugget. The stage was clearly set not only for transmutation but also multiplication of precious metals.

Alchemy around Alexandria

Alchemy doubtless prospered in the area around Alexandria, but very little survives to record its progress. Works that could have given an insight into the origins of alchemy were probably destroyed in AD292 when the Roman emperor Diocletian demanded the burning there of all 'books written by the Egyptians on the *cheimeia* of silver and gold'.

The earliest text we have, besides the supposed transcription of the Emerald Tablet, is a fragment of a work by the Greek-Egyptian alchemist and mystic Zosimos of Panopolis. (Panopolis is now Akhmim.) It was written around AD300, but is preserved in a later copy. His work is clearly distinct from the recipes of the papyri. His purpose was the transmutation (not imitation) of metals, and he sought to establish principles and theories and proceed through a logical process. He considered that 'vapours' carried

A recreation of how an Alexandrian chemical workshop might have looked in the 1st century AD, based on archaeological remains and contemporary records.

characteristics of matter and that the solid part or body of matter was generic. It then followed that if he separated the vapours and the body, he could change the nature of a lump of matter by combining it with different vapours. Most of his methods used heat as a way of separating and combining body and vapours.

LADIES FIRST

Zosimos credited four women with the invention of alchemy, two of whom are named: Mary (or Miriam) the Jewess, and Cleopatra the Alchemist. Most of Zosimos's own writings have been lost, but the surviving fragments feature instruction to a female pupil. This might have been a literary device, or he might really have had such a pupil, but either way it is interesting that the pupil was female. Women evidently played a significant part in early chemistry, or at least in the culture surrounding it.

Starting with secrets

Alchemy is traditionally a secret art. At least one early collection of chemical recipes emphasized secrecy, but that was because it made good business sense to hide the methods by which the craftsman made money.

Zosimos also stressed the need for secrecy. As his aim was the transmutation of base metals into gold, he didn't want just anyone finding out how to do it or the value of gold and of his skills would plunge. With his writings began a tradition of secrecy, cryptic allusion, allegory, metaphor, symbols and codes to disguise the true nature of the chemistry being described. This practice would continue throughout the coming centuries of alchemical endeavour. For example, mercury is described, rather than named, as 'the silvery water, the hermaphrodite, that which flees without ceasing' and so on, and some of his processes are presented in the form of allegorical

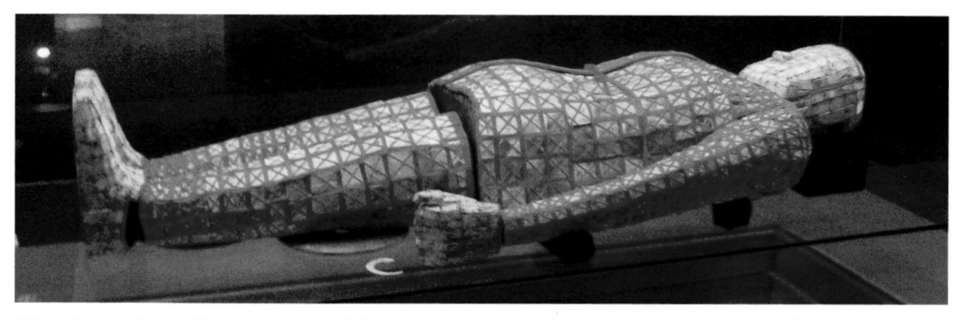

The jade burial suit of King Zhao Mo (died 122 BC) might have been intended to confer eternal life, since jade, gold and cinnabar were supposed to be agents of immortality.

dream sequences in which metal men are dismembered on an altar. There had to be some kind of logic to the choice of code words and some method to the allegory as the idea was to communicate with the initiated, not just to hide information from the casual reader.

Fragments of Zosimos's work exist in the original Greek and in translations into Syriac and Arabic. His book ran to 28 volumes, so there was clearly a lot to say about alchemy by then. He refers to previous writings (now lost) and practical techniques which show that much apparatus had already been developed, adapted from utensils used by artisans and cooks. Part of only one volume survives, but even from this it is clear that he had apparatus for distillation, filtration and sublimation (techniques described elsewhere in this book), a water-bath for gentle heating and various receptacles and ovens.

Alchemy in China

Alchemy seems to have developed pretty much concurrently in the West and in China and it is certainly possible that there was communication between the two communities, as trade routes were operating between China and Alexandria in the early centuries AD. The earliest mention of alchemy in China is an edict from 144BC prohibiting the practice, saying that those who made counterfeit gold would be executed. About AD180, a commentator remarked that alchemy had previously been allowed, but alchemists had wasted their effort and money on making fake gold, then had to turn to crime to alleviate their poverty, and that was the reason for banning the practice. Even so, in 133BC – only eleven years after the edict – the emperor Wu indulged an alchemist who promised to make him immortal. A later emperor, in 60BC, employed an alchemist who claimed he could turn base metal into gold. That emperor then condemned the alchemist to death when he failed (but ultimately let him off the sentence).

The features of immortal life and the transmutation of metals into gold were thus evident in Chinese alchemy at about the same time as the West developed an interest in transmutation. A text written in the 3rd century AD, in the Taoist Wei Boyang's *Tsan Tung Chi*, refers to the 'pill of immortality',

made of gold. The extraordinarily flamboyant style makes it difficult to extract any useful instructions from the book (see box, below), but just in case anyone did manage, it is not that straightforward. An alchemical treatise written by Go Hung in the late 3rd or early 4th century AD stresses that it's not just a matter of doing the right things: the alchemist must have learned from an expert who has passed on secrets which are only communicated orally, must worship the right gods, be specially blessed, have been born under suitable stars, must purify himself with perfumes, have fasted for 100 days(!) and should carry out the work on a 'famous great mountain, for even a small mountain is inadequate'. This aspect of special circumstances and of being 'favoured', while particularly prominent in the Chinese tradition, was certainly not absent from Western alchemy. It's not possible to work out exactly what all the ingredients were in Go Hung's 'Divine Elixir', which is a shame since it could convert mercury or a lead–tin alloy into gold or, if taken for 100 days, would confer immortality.

An 18th-century depiction of Tao Hongjing (AD451–536), pounding rocks to make an elixir of immortality following the recipe of Go Hung.

Practising chemists

Alchemy borrowed from the expertise and apparatus of artisans skilled in producing

> *'Above, cooking and distillation take place in the cauldron; below, blazes the roaring flame. Afore goes the White Tiger leading the way; following comes the Grey Dragon. The fluttering Scarlet Bird flies the five colours. Encountering ensnaring nets, it is helpless and immovably pressed on, and cries with pathos, like a child after its mother. Willy-nilly it is put into the cauldron of hot fluid to the detriment of its feathers. Before half the time has passed, Dragons appear with rapidity and in great number. The five dazzling colours change incessantly. Turbulently boils the fluid in the cauldron. One after another they appear to form an array as irregular as dog's teeth. Stalagmites, which are like midwinter icicles, are spat out horizontally and vertically. Rocky heights of no apparent regularity make their appearance supporting one another. When Yin and Yan are properly matched, tranquility prevails.'*
>
> Wei Boyang, *Tsan Tung Chi*, 3rd century AD

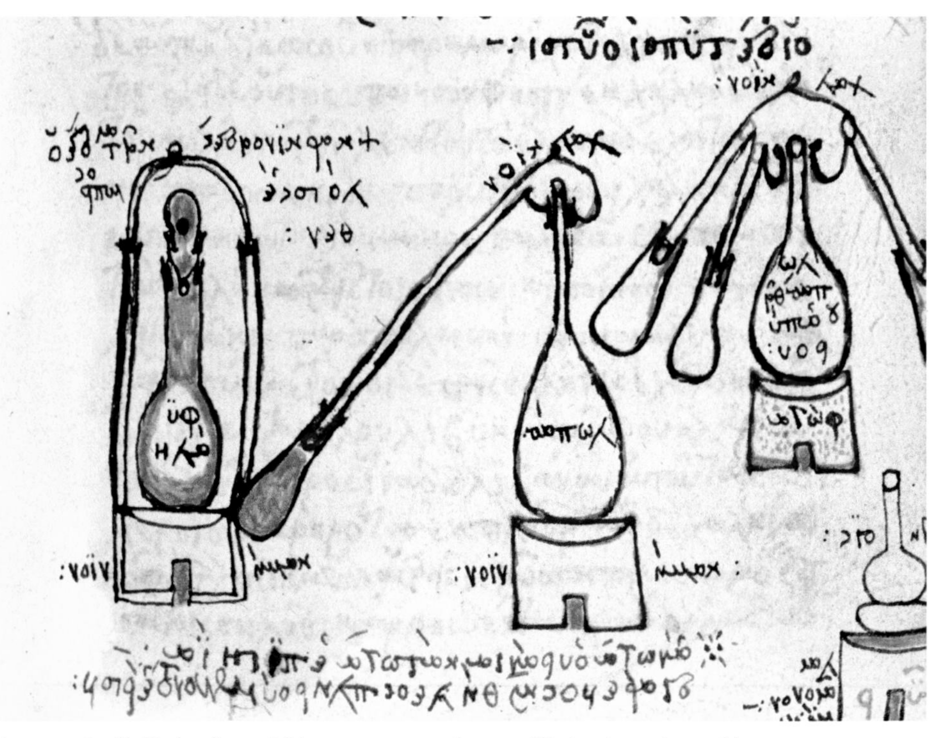

Apparatus for distillation, from a 16th-century manuscript copy of Zosimos's treatise on alchemy.

STILL USING STILLS

The principal piece of equipment used in distillation is the still or alembic, in which liquids are heated. The alembic has three parts: a container for the original liquid, called the 'cucurbit'; the 'head', which forms a cap over the cucurbit; and one or more receivers connected by tubes to the head. The liquid in the cucurbit is heated and vapours rise from it to condense on the inside of the head and run down the collecting tubes into receivers.

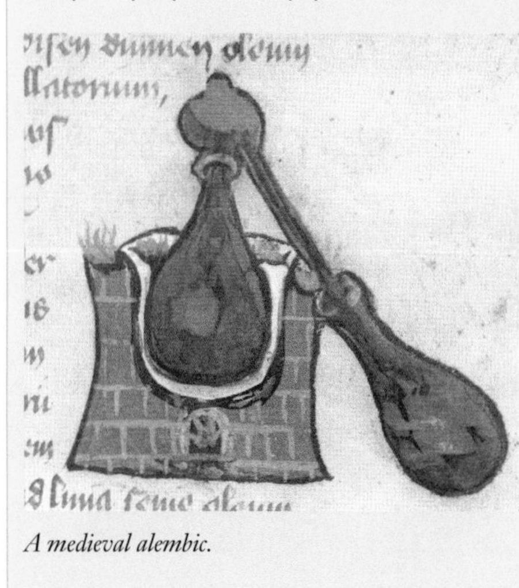

A medieval alembic.

According to Zosimos, the alembic was invented by Mary the Jewess, but it has also been attributed to Cleopatra the Alchemist. It was used to distil alcohol and to separate plant oils for making perfumes and medicines. The same principle is still used in alcohol distilleries.

dyes, pigments, glass, ceramics, medicines and perfumes, and in metalworking and metallurgy. The Egyptian alchemists adapted and improved both techniques and equipment to suit their own ends. Their principal items of apparatus were furnaces (and bellows to increase their temperature), crucibles and alembics. Their methods mostly centred on heating substances, either alone (which naturally meant in air, so in the presence of oxygen) or in combination. Zosimos was particularly interested in heating substances in the presence of a vapour.

Processes included the distillation of liquids and purifying metals by melting them in a perforated crucible and collecting the liquid that dripped through. The alchemists sublimated substances by heating them and then cooling the vapour rapidly, collecting the condensed solid. Calcination (forming a calx, or oxide) was effected by heating a substance so that it combined with oxygen from the air. The same process had been used in smelting metals in prehistoric times, but now it entered the laboratory as a deliberate step in its own right.

Zosimos was significant in bringing together practical chemistry and theorizing about the nature of matter. The Ancient Greeks who had thought about the changing of matter had been theorists and philosophers, and the artisans had been practical, hands-on people who had no need of a theoretical framework for their work. Fragments of other alchemical works surviving from slightly later in Hellenic Egypt seem to mark a return to theory, with more mysticism and secrecy and less of the practical chemistry at which Zosimos was apparently adept.

Basic matter

Olympiodoros, who left a commentary on Zosimos in the 6th century, is notable in developing Zosimos's distinction between vapour and body and Thales's search for a single fundamental type of matter. He proposed a common 'matter of metals', which is the substrate on which all metals are built; the distinct types are produced by different qualities being given to the basic 'matter'. The logical conclusion is that if the qualities of base metal can be stripped away to reveal the underlying 'matter', this can then be imbued with the qualities of gold or silver, so effecting a transmutation. This would become the basis of the alchemy practised by the Arabs and later in Europe.

The philosophers' stone

The 'philosophers' stone' became the ultimate goal of the alchemists. It was thought to be a physical substance capable of effecting seemingly miraculous transformations. A small portion of it cast into base metal could transform the metal into silver or gold; consumed, it could restore health and extend life; in the right recipe, it could animate a homunculus (miniature person – see page 64). But the philosophers' stone was a powerful metaphor as well as a supposedly real, though elusive, material and was spoken of in regard to the purification of the soul, the ultimate spiritual transformation.

The concept can be traced back to Zosimos. He referred to 'water of sulphur' as a transmuting agent, and to the 'stone that is no stone'. (He also used the term 'water of sulphur' for other things; it was common among alchemical writers to call

one thing by many names and use one name for many things.) The transformative agent was later referred to as an elixir. In whatever form it took, stone or elixir, the transformative agent became the elusive goal of all alchemy.

At the House of Wisdom in Baghdad, from the 9th to the 13th centuries, the learning of the Greeks was translated into Arabic and augmented by the original work of Islamic scholars. Alchemy was one of many subjects treated here.

Arab alchemy

In 640, Alexandria fell to Arab invaders and was annexed to the Islamic states. Soon after, translators set to work – first in Damascus and, from 762, in Baghdad – rendering all kinds of Greek texts into Arabic. They effectively transferred Greek intellectual culture wholesale into the new and burgeoning Arab civilization.

Little detail is known about how interest in alchemy was kindled among Islamic scholars and scientists, but one historian of the 10th century relates (not necessarily reliably) that sometime between 754 and 775 al-Mansur's ambassador visited the Byzantine emperor, who demonstrated a spectacular alchemical feat. He threw a white powder into a vat of molten lead, which promptly turned to silver, and a red powder into a vat of molten copper, turning it to gold. On hearing of this, al-Mansur is said to have commanded the translation of alchemical treatises from Greek. Translation of scientific and medical works certainly accelerated and flourished under al-Mansur, and in later lore the philosophers' stone was often said to come in two forms, white and red, with different transformative capabilities.

MORE MYTHICAL ORIGINS

Arabic alchemy also has a legend concerning its origins, although it's not quite as exotic as the tale of the Emerald Tablet prised from the dead hands of Hermes by Alexander the Great. It concerns a young Umayyad prince, Khalid, dispossessed of his caliphate by a relative who offered to hold it as protector for him until he reached maturity, but who then married the prince's mother and made his own children heirs. Khalid's mother killed her new husband.

It clearly wasn't a very safe or enjoyable environment for the young man so he understandably left his homeland, fleeing to Egypt to start a new life studying the Greek intellectual legacy. There he met a Christian alchemist called Marianos who, hoping to convert the Muslim prince, shared with him the secrets of alchemy, including how to make the philosophers' stone. Khalid went on to write alchemical texts of his own. It's a nice story, but the works attributed to Khalid were written more than 100 years after he actually lived.

A way forward

The first great Arab alchemist was Jabir ibn Hayyan, known in the West as Geber. He has even been called the 'father of alchemy' as he left the first comprehensible, systematic account of alchemy and turned attention back towards practical chemistry rather than esoteric spiritual ponderings. Or so tradition has it. It's not actually clear whether he existed or whether his identity is a composite of several people. He was supposed to have lived in the 8th century, but scholarship has shown that 'his' works show influences that date from around a century later, including the rise of Shi'ite philosophy. Doubts about his existence date from the 10th century. Around 3,000 works were attributed to him (though they were mostly very short), and it seems more likely that Jabir was a name attached to works to give them the mark of authority, no matter who wrote them. For convenience, we will treat him as the author of the text ascribed to him.

A central idea of later alchemy is stated by Jabir: that all metals are a mix of mercury and sulphur in different proportions. The

Essential elements of alchemical theory, mercury (top) and sulphur. Mercury is the only metal liquid at room temperature, a feature which marks it as special.

> *'The first essential in chemistry is that you should perform practical work and conduct experiments, for he who performs not practical work nor makes experiments will never attain to the least degree of mastery.'*
>
> Jabir

perfect proportions produce gold (of course). To produce gold, though, the sulphur and mercury have to be of impeccable purity. Any imperfections will result in different metals. If the ingredients can be purified and the proportions successfully adjusted in a quantity of any other metal, it will, naturally, become gold. Thus Jabir set out what appears to be a logical reason why alchemical transmutations should be possible, made an excuse for failure (impurities), and set in train around 1,000 years of fruitless attempts to make them happen.

Theory held that metals formed naturally within the earth might change from one to another over a period of time as a result of natural processes. Over hundreds or thousands of years, heat and washing by water percolating through the ground might refine base metals, turning them to gold and silver. This did not seem unreasonable, as ores often contain a mix of metals – perhaps one was slowly morphing into the other. What the alchemist was attempting, then, was simply replicating and accelerating a process that could happen naturally.

Four natures

It's not quite as simple as it sounds, though. Jabir believed there were four elementary natures, based on Aristotle's ideas about matter: heat, coldness, dryness

> *'You should know that the mineral bodies are vapours which are thickened and coagulated according to the working of nature over a long time. What is first coagulated in them is quicksilver and sulphur. And these two are the origins of the mineral. . . . A temperate concoction has remained with them with heat and humidity until they are congealed, and from them the [mineral] bodies are generated. Then they are gradually mutated until they become silver and gold in a thousand years.'*
>
> Pseudo-Razi, *De aluminibus et salibus* (*On Alums and Salts*), translated by John Norris, 2006

and moistness. By repeated distillation, he thought it was possible to produce matter that instead of having two qualities had only one. So if you took water and removed all the 'wet' from it, you ended up with a white crystalline or powdery solid that was matter with 'cold' as its only quality. Clearly, if you could isolate each quality, you could then add it as necessary to adjust the qualities in any other matter. Simplifying rather, sulphur provided the hot and dry parts of a metal's nature and mercury provided the cold and moist parts.

Of course, to adjust the qualities in a sample to turn it into gold, it was necessary to know how much of each to add, and that meant knowing the composition of the starting material as well as the required composition. Here Jabir strayed from the practical, empirical path. He employed a complex numerological scheme that worked with the Arabic spelling of the name of the metal to reveal the proportions of the four

natures to be found in it. He went further, subdividing the four natures into seven intensities, giving 28 categories altogether. A good deal of mathematical jiggery-pokery was involved, and we need not go into the detail. The result was that in order to transmute a metal, it was necessary to calculate the ratios of natures that needed to be added and work out from the mass of metal to be transformed how much of a special transformative substance called an elixir would be required. In Jabir's time, there were seven recognized metals: gold, silver, lead, copper, iron, tin, mercury – which was not always classed as a metal – and something known as *khar sini*, which was used to make up the number to seven when mercury wasn't included. It seems to have been an alloy of copper, zinc and nickel, sometimes now called 'nickel silver'. Of these, two were noble (gold and silver) and the other five were base, so there were potentially ten types of transformation that could be achieved with appropriate elixirs.

It follows that there were specific elixirs to effect every possible type of transmutation, as each of the base metals had different proportions of each nature. In addition, there was a grand or master elixir, that could effect any type of transmutation. It was this superlative elixir that was also known as the philosophers' stone. Only the most dedicated and hardworking of alchemists would go to sufficient trouble, and have sufficient skill, to produce the best elixir.

Getting down to the nitty-gritty

Muhammad ibn Zakariya al-Razi, known as Rhazes in the West (854–925), is known principally as a physician, and wrote some important medical treatises, but was also an active and practical alchemist. He seems to have been far more interested in experimentation and practical chemistry than in the mystical aspects of alchemy.

Al-Razi had a well-equipped laboratory and listed the equipment an alchemist should have, dividing it into those items used for dissolving or melting metals and those used in transmutation. The first group included tools for making a fire and handling hot objects, including bellows, crucibles and tongs, and items for breaking up the material to be melted or dissolved, such as pestle and mortar, stirring rods and shears. The second group included the equipment necessary for distilling, including furnaces and ovens, retorts, alembics, water bath, various items of glassware, sieves and filters.

He classified chemicals into six groups:

- Four spirits: mercury, sal ammoniac (ammonium chloride), sulphur and arsenic sulphide.
- Seven 'bodies', or metals: silver, gold, copper, iron, 'black lead' (graphite), *khar sini* and tin.
- Thirteen stones: marcasite (iron sulphide), magnesia or periclase (magnesium oxide), malachite (copper carbonate hydroxide), zinc oxide, talcum, lapis lazuli, gypsum, azurite, haematite (iron oxide), arsenic oxide, mica, asbestos and glass.
- Seven vitriols (sulphates); sulphuric acid could be obtained from vitriol, and this was used in transmutations as a solvent for metals.
- Seven borates, including natron.
- Eleven salts, including common salt (sodium chloride), ashes, naphtha, lime (calcium oxide) and urine.

BRANCHING OUT

Before Jabir, alchemists had generally restricted themselves to working with minerals and metals, but Jabir listed organic ingredients useful to the alchemist, including the blood, hair, bones, marrow and urine of lions, foxes, cattle, donkeys, gazelles and vipers, and plant material from pears, olives, onions, ginger, aconite, pepper, mustard, jasmine, love-in-a-mist and anemones. This seemed quite sensible – from an alchemical point of view – as it's much easier to break down organic matter than some of the metals and minerals an alchemist might choose to work with in trying to extract pure, single natures to use in an elixir.

His books give procedures for various reactions, all rigorously documented without the flowery allegorical language characteristic of much alchemical writing.

Al-Razi accepted the theory that all metals are made of mercury and sulphur, but maintained that some metals also contain a salt of some kind. He rejected Jabir's complex numerological method of working out how to correct the balance to produce gold, but he was not set against transmutation. Indeed, he extended the remit of alchemy to include also the transmutation of stones, rock crystal and glass into gems.

Transmutation explained . . .

Al-Razi gave a comprehensive and comprehensible account of the complex series of processes involved in transmuting base metals into gold. Those he described

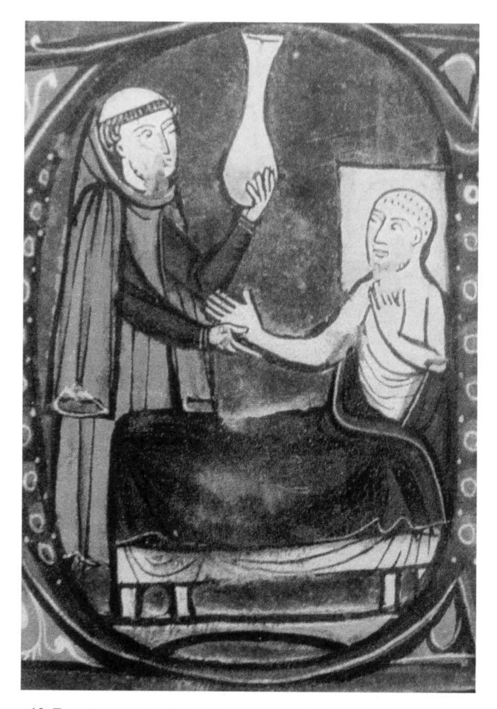

Al-Razi, seen in his role as medical practitioner, examining a patient's (rather large) urine sample.

AL-IKSIR

The word 'elixir' comes from the Arab *al-iksir*, itself formed from the Greek *xerion*, a medicinal powder used to heal wounds. The ending 'ion' is grammatical and can be removed; 'al' is the Arabic article. The name reflects the link with medicine that was also based around the qualities of hot/cold and moist/dry, expressed in the form of the four humours that must be correctly balanced for the body to be healthy. The use of 'al-iksir' suggests that matter is healed and brought to a perfectly balanced state through the corrective action of the substance, and medical metaphors were certainly common in alchemical writings.

include distillation, calcination (formation of a calx or oxide), solution, evaporation, crystallization, sublimation, filtration, amalgamation and ceration (softening a hard substance into a pliable paste, usually by heating and mixing with liquid). To effect a transmutation, the steps were:

- to purify the substances to be used through distillation or other appropriate methods
- using ceration, to convert the substances to a soft consistency which, when dropped on a hotplate, melts easily, producing no fumes
- to dissolve the resultant paste in 'sharp waters' (usually alkalis or liquids containing ammonia)
- to mix together the solutions
- to coagulate or solidify the mixture, perhaps by evaporation.

Arab distillation equipment from the 10th to 12th centuries.

The substances were chosen according to the proportions of the different properties needed. The final product, solidified at the end, would be an elixir. The process is even harder than it sounds, as some stages would be repeated many times – even hundreds of times – to gain as pure a product as possible.

If (when) it didn't work, the alchemist would explore the product he or she had achieved to discover its properties. This was a gift to the history of chemistry, as alchemists set about making pure versions of all kinds of unknown compounds and then testing them to see what they did. Some compounds were found to have therapeutic value – although there were probably some unfortunate outcomes along the way as random substances were tested on hapless patients.

Useful results

Part of Jabir's legacy was the meticulous recording and investigation of chemical reactions and their products. This led quickly to some important discoveries. In the 10th century, for instance, Abu Mansur clearly distinguished between potassium carbonate and sodium carbonate; discovered plaster of Paris and found its application in setting broken bones; described arsenic oxide and silicic acid; found that antimony has a metallic sheen when first cut but quickly dims as it oxidizes, and that copper heated in air forms copper oxide that can be used as a black hair dye. Some of the

products of alchemical experimentation, generally inorganic, were the precursors of the iatrochemistry (medicinal treatments originating in alchemy) practised and popularized by Paracelsus in the 16th century (see page 60), and ultimately of our own chemotherapies.

Transmutation denied

Abu Ali ibn Sina (Avicenna in the West) was one of the most brilliant Arab physicians and scientists of the 11th century. He was also a dissenter with regard to alchemy. Ibn Sina agreed with Jabir in principle on the composition of metals, but denied that alchemists could ever achieve transmutation. He accepted that they might produce substances that looked very like real gold and silver, but they would only ever be imitations. This, he explained, was because 'art is weaker than nature' and conversion between forms was impossible:

'I regard [transmutation] as impossible, since there is no way of splitting up one metallic combination into another. Those properties that are perceived by the senses are probably not the differences which distinguish one metallic species from another, but rather accidents or consequences, the essential specific differences being unknown.'

This astonishingly insightful warning would eventually turn out to be correct: the fundamental differences between metals are indeed not discernible by the senses, being a matter of atomic configuration, and the

An Arab sage (on the left), consulting an alchemical tablet.

differences manifest are consequences of those fundamental differences.

Ibn Sina's good sense did not prevail; others defended alchemy against his attack and the search for elixirs continued unabated. But the focus of alchemical activity was about to shift. Much of the subsequent Arab work involved a return to earlier texts. There were also some notable exceptions, though, particularly the work of Maslama al-Majriti. His book *The Sage's Step* includes a detailed procedure for purifying gold, and instructions for preparing mercuric oxide that pay more attention to the quantities of both products and reactants than would be common again before the 18th century. But this was not typical. The general trend was of stagnation in the Arab lands as the action switched to a new stage: Europe.

Alchemy enters Europe

The dawn of alchemy in Europe can be dated very precisely to Friday 11 February 1144. On that day, the Englishman Robert of Chester completed his translation of the Arab *Book of the Composition of Alchemy* into Latin. He, along with other European scholars, had been welcomed in southern Spain, still largely an Arab territory, to study and translate the great works of Arab scholarship and of the Greek legacy. It was through Spain, particularly Cordoba and Seville, that the intellectual inheritance of Classical Greece, Graeco-Roman Egypt, Syria and all the Arab nations entered Europe and became the foundation of European learning.

The *Book of the Composition of Alchemy* is the text that was supposedly written by Marianos for Khalid ibn Yazid explaining

Ibn Sina employed chemical knowledge in the service of pharmacology. Paracelsus would do the same thing in Europe 500 years later.

the secrets of alchemy, so essentially putting European alchemy on the same footing as Arab alchemy. This was not the only text translated. Over the coming years, a host of talented translators worked to render into Latin – sometimes by way of Castilian Spanish – the alchemical and other works of the Arabs, including texts by Jabir, al-Razi and ibn Sina. Before long, European scientists were adding their own investigations to the corpus.

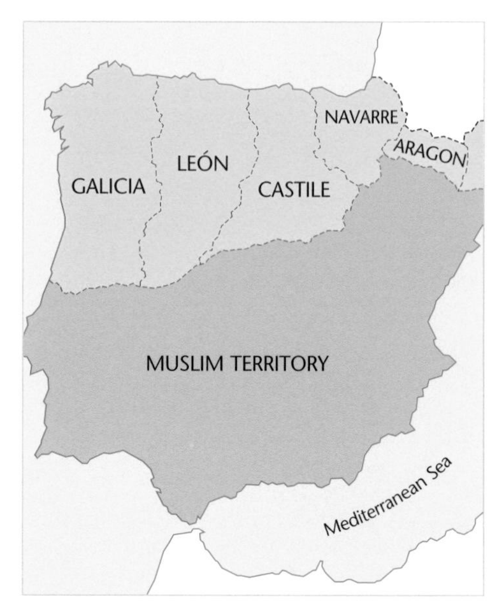

Spain was divided between the Spanish kingdoms and Muslim territories in 1065.

Fake Arab

Just as Arab writers sometimes appropriated the identity of their Greek predecessors to lend authority to their works, so some of the early European alchemical writers appropriated Arab identity. One of the most influential and widely used texts of the

The city of Cordoba in Moorish Spain was one of the great intellectual centres through which Arab learning passed into medieval Europe.

A STICKY END

The great intellectual centre of Baghdad and the Islamic Golden Age came to a terrible end in 1258. Mongolian invaders led by Hulagu (grandson of Genghis Khan) destroyed the libraries, mosques, schools and hospitals of the city, throwing the books into the River Tigris and slaughtering the scholars. The river was said to run black with ink and red with blood for days.

TAKE ONE BASILISK . . .

An early European alchemical recipe survives in an otherwise quite down-to-earth and practical chemical recipe book produced by Theophilus around 1125. It explains that Spanish gold can be made from a mixture of red copper, basilisk powder, vinegar and human blood. It would have been difficult to refute Theophilus's claims for his recipe.

Getting hold of basilisk powder isn't easy. You need to start by overfeeding cockerels until they mate and produce an egg, then persuade a toad to sit on the egg until it hatches. Finally, raise the basilisk underground in a kettle and burn it to yield the powder.

The basilisk is variously described as a crested snake or cockerel with a snake's tail.

Middle Ages was the 13th-century *Summa perfectionis* (*The Sum of Perfection*) by an author who claimed the name Geber (the Latinized form of Jabir). He is sometimes known as Pseudo-Geber, and might have been the Italian friar Paul of Taranto. His work shows heavy Arab influence, but is more practically oriented, being clearly based on laboratory experience. Like the original Jabir, Geber describes three grades of elixir, but differs from his predecessor in allowing only mineral sources for preparing the elixirs. He gives information about how to purify substances and also how to test the purity of the alchemical products – assays for gold and silver (see page 171).

On the nature of matter, Geber combines the mercury–sulphur theory about metals with an idea drawn from Aristotle that solid substances have parts and pores. In gold, the parts are very densely packed, making gold dense and incapable of disintegration. In base metals, the parts are less densely packed and the pores make the metal softer or more capable of being broken apart by heat – the particles of fire can worm their way in between the parts of metal and break them asunder. Although Aristotle was not an atomist, he did believe there was a lower limit to the size of a piece of matter that still retained the properties of that substance. So beyond a certain point, cutting a piece of gold in half would yield something too small to have the properties of gold. (We might agree in some ways: one molecule of water, while still water, can't be said to be wet even though it has the properties and potential to create wetness en masse.)

Summa gives clear instructions for the preparation of acids and for laboratory procedures, explains the mercury–sulphur composition of metals and the principles of transmutation and defends alchemy against criticism. Its clarity no doubt contributed to its success; it was one of the most widely read and influential of the medieval alchemical texts.

Albertus Magnus: scientist or sorcerer?

With the German cleric, scholar and scientist Albertus Magnus we enter the realm more commonly associated with alchemy in the popular imagination. Much has been credited to him, and legends have accreted around him that include acts of magic (see box, opposite).

Albertus was very much a hands-on chemist. Perhaps as a consequence of that, some of his contemporaries claimed that he was in communication with the devil and worked magic. Legend tells that he successfully created the philosophers' stone and passed it on to Thomas Aquinas (1225–74), but Aquinas destroyed it as he suspected it to be the work of the devil. Albertus is credited with identifying arsenic around 1250 – the only element discovered between ancient times and the start of the chemical revolution in the 17th century. Although arsenic had been in use since the Bronze Age, it is not known to have been isolated before.

Mercury was considered to be 'water of the Sun and Moon', shown in this 15th century manuscript as two entwined trees producing the principle of mercury in the centre.

Albertus accepted the prevailing view that moist and dry vapours worked in the earth to produce metals and minerals, and that mercury and sulphur were the essential ingredients of all metals. Picking up on ibn Sina's contention that matter produced by chemistry is never exactly the same as that produced naturally, he maintained that alchemical gold is the same as natural gold in all regards except

two (and they are related): a wound made with a weapon forged from real gold will not fester, while if the weapon is made of alchemical gold it will; and alchemical gold does not share the medicinal qualities of natural gold. Alchemical iron, he asserted, differs from natural iron in not responding to a magnet. Albertus also claimed that when he tested alchemical gold, it survived six or seven cycles of firing but then fell apart or was consumed by the fire.

Alchemy and the Christian church

Albertus Magnus, his student Thomas Aquinas, and the English scholar Roger Bacon (*c.*1219–*c.*1292) were instrumental in drawing Aristotle's work into the Christian Middle Ages. They accommodated it to the requirements of Christianity so that it could be respectably studied, and set a pattern of deference to Aristotle's authority

Albertus Magnus lecturing to students.

that ultimately blighted scientific progress. This included drawing alchemy into the Christian fold.

Albertus and Bacon provided foci for the consolidation of learning rather than providing much in the way of new scientific knowledge. This was a vital role, and one that was supplied by several encyclopaedists from the early Middle Ages onwards. Bacon dealt with practical alchemy, explaining how to make things by chemical means, including gold, medicines and gunpowder (though the

ALBERTUS MAGNUS (c.1193–1280)

Born in Bavaria sometime between 1193 and 1206, Albertus Magnus (Albert de Groot) was the son of the Count of Bollstädt. He appeared fairly stupid as a boy, but showed an interest in religion. One night, it's claimed, the Blessed Virgin Mary appeared to him in a vision and after that his intellect was considerably sharpened. He went on to become a great scholar, in particular translating and commenting on the works of Aristotle. He was highly regarded for his powers of logical and systematic thought, and wrote treatises on alchemy and other topics. Appointed bishop of Ratisbon in 1260, he resigned after three years to concentrate on his scientific investigations.

> *'Natural science does not consist in ratifying what others have said, but in seeking the causes of phenomena.'*
>
> Albertus Magnus, 13th century

proportions he gave would not have made good gunpowder). He emphasized that the practical aspects could provide evidence of what lay behind the behaviour of matter, though he felt this was not fully accessible, and certainly not to the (pagan) ancients. In keeping with this, he promoted Aristotle's principle of making many and detailed observations before attempting to deduce scientific proofs, not only in the realm of alchemy but of all science. This made him an early proponent of something like the scientific method (see page 71).

Bacon and Albertus together were said to have created a bronze head which could answer questions. Needless to say, there is no evidence that they actually achieved such an astonishing feat. There is much legend about and little confirmation of alchemical practice by both figures. In reality, they summarized and clarified some alchemical ideas but made little or no new progress in practical chemistry. They probably did not write some of the works attributed to them. Bacon was a thorough believer in the possibilities of alchemy; among the texts he respected was the *Liber Secretus* (*Secret Book*), supposedly written by Artephius, an Arab scholar working around 1150. In his book, Artephius claims to have been born in the 1st or 2nd century, having lived to be more than 1,000 years old by dint of making the philosophers' stone and using it to extend his life.

Beyond metals

Alchemy as it entered Europe was concerned mainly with the transmutation of metals, but the subject would itself undergo transformation in the coming centuries. Its remit broadened to include physical and even spiritual health, and it became explicitly Christianized. Even so, it would never be something the Church was entirely comfortable with.

Alchemy and the soul

The alchemists of the 13th and 14th centuries further helped to forge links between Christian theology and alchemy. By around 1300, a Christianized version of the microcosm/macrocosm alchemical harmony existed. It was taken that just as suitable purification methods could be used to perfect base metals, turning them into gold, similar principles applied to perfecting the human soul (and body). This was easily accommodated in a worldview that saw the whole of Creation loaded with messages for humankind and geared entirely towards the salvation of human souls. Everything in the macrocosm – the

> *'To say that there is a soul in stones simply in order to account for their production is unsatisfactory: for their production is not like the reproduction of living plants, and of animals which have senses. For all these we see reproducing their own species from their own seeds; and a stone does not do this at all. We never see stones reproduced from stones . . . because a stone seems to have no reproductive power at all.'*
>
> Albertus Magnus, 13th century

THE SECRET OF SECRETS

Bacon's name is often linked with a text known as the *Secretum* or *Secreta Secretorum* (*The Secret of Secrets*). It claims to be a letter written by Aristotle to Alexander the Great (his pupil) but was probably written in Arabic in the 10th century and translated into Latin in the 12th. It covers a number of topics, including alchemy and other sciences. Bacon produced his own annotated edition of *Secretum* as well as citing it frequently. It was among the most widely read books of the 12th and 13th centuries.

universe at large – was seen to reflect the state of the microcosm and hold lessons for those willing and able to find them.

When the steps in alchemical practice are presented in allegorical form there is plenty of opportunity for confusion. The *Metaphorical Treatise*, a text attributed to Arnauld de Villanova (*c*.1240–1311), likens the stages of the preparation of the philosophers' stone from mercury to the stages of the life of Christ. His account of the process is entirely obfuscated by the allegory. Yet this need not be the case. A slightly later writer, Jean de Roquetaillade (*c*.1310–*c*.1366), gave clear, sequential steps describing the chemical processes and ingredients, yet still cloaked them in metaphors. The presentation served to hide the information, but the information was still there and could be uncovered by the initiated. Whether lucid or obscure, this type of allegory forged a link with Christian theology that would continue through subsequent centuries, elevating alchemy and perhaps helping to protect it from persecution.

A page from an early 15th-century manuscript of the Secreta Secretorum; *the top image shows the building of the Tower of Babel and the bottom shows Zoroaster with two demons.*

Alchemy and the body

De Roquetaillade also developed alchemy's links with medicine and the prolonging of life. This had practical applications: he learned about distilling alcohol, producing *aqua vitae* (ethanol in water) that he then used to extract active ingredients from medicinal herbs and plants. In most cases, he found it more effective than water.

It was a short step from using *aqua vitae* medicinally to incorporating it in recipes for the philosophers' stone. The step was soon taken, in a book named *Testamentum* that appeared in 1322. It concerned itself with the transmutation of metals, the forming of gems and the preservation of health, and attributed all three powers to the philosophers' stone, a universal healer. The application of alchemy to issues of health would be most thoroughly investigated 200 years later by the maverick Swiss physician Paracelsus (see page 60).

Fake alchemy and fake alchemists

It goes without saying that attempts to make gold from lead and to discover an elixir of youth or health were unsuccessful. But there was a good profit to be made from pretending to succeed. During the European Middle Ages, charlatans who claimed to be able to achieve the goals of the alchemists gave bona fide alchemists a bad name. Some countries passed laws banning transmutation, or at least banning making money out of promises of transmutation. Literary authors of the period such as Dante Alighieri in Italy and Geoffrey Chaucer in England criticized or mocked scammers masquerading as alchemists. Even so, this didn't mean that people no longer believed transmutation was possible or thought that

MICROCOSM AND MACROCOSM

Alchemy proposed harmony between the macrocosm and the microcosm, and parallels between what is 'above' and 'below' are mentioned in the Emerald Tablet. This extended to astronomical/astrological correspondences. The metals were already associated with heavenly bodies, with gold linked to the Sun, silver to the Moon, and the base metals each assigned a planet. The association of the Moon with silver and the Sun with gold are fairly obvious. Mars was probably associated with iron because iron was widely used for armour and weapons (and Mars was the god of war), and Venus was associated with copper because both were linked with the island of Cyprus (home of Venus and source of copper ore).

The human represents the microcosm, in harmony with the macrocosmic universe.

the philosophers' stone might not exist, just that it was a business prone to exploitation. Although Henry IV of England legislated against false alchemists multiplying metals, it was still possible to buy a licence to practise alchemy to attempt to transmute substances into gold.

The alchemical quandary

For the less sceptical European elite, alchemy posed a dilemma. Its promise of the transmutation of base metals into gold seemingly offered a route to great wealth if it could be harnessed by the right people. On the other hand, it also threatened financial

Chemists preparing aqua vitae *using distilling apparatus.*

In Dante's Inferno *(Hell), 'falsifiers', including alchemists, are punished with a terrible affliction that causes them to scratch the skin from their bodies.*

ruin and degradation of the currency in the hands of the wrong people. It's clear that if a method of making gold from any old thing were to get out, the value of gold would plummet. The inevitable outcome was that alchemy was banned, but at the same time some rulers secretly supported or entertained alchemists, hoping to benefit personally from any discovery.

There was even an element of consumer protection in the legislation. Greedy merchants could easily be hoodwinked by charlatans claiming to be alchemists who would cheat them out of a supply of gold, silver or gems with promises of multiplying it – promises that, predictably, amounted to nothing. Alchemy became a crime punishable by death across Christendom. Even so, legend has it that in the early 14th century Pope John XXII practised alchemy himself and claimed to have successfully increased the Vatican's wealth in the process.

Inevitably, any suspicion of success brought attention, often unwanted. There are plenty of tales of alchemists being abused, imprisoned, tortured or pursued because someone or other – usually a hopeful monarch – believed they had acquired the philosophers' stone and could make gold or prolong life but then found that the alleged alchemist was not compliant or competent.

HOW TO MAKE A LIVING – OR A KILLING – AS A CHARLATAN ALCHEMIST

Here are four ways to turn a profit as a fake alchemist:

- In place of a solid stirring rod, use a hollow lead rod that you have filled with gold powder, the end sealed with wax. As you stir the heating mixture in your crucible, the wax will melt and the gold will mingle with your mixture.
- Make a nail, the top half of which is iron and the bottom half gold. Cover the gold with black paint. Ensure the level of mixture in your flask or crucible comes to halfway up the nail, to the join. Stir a suitable paint-dissolving mixture using the nail, or dip the nail into the mixture; the part of the nail that has contact with the mixture will appear to turn to gold as the paint dissolves.
- Make a coin from a white alloy of silver and gold (not too much silver). When you dip the coin in acid, the silver will dissolve, making it appear that the coin has been changed into gold.
- Heat copper in the presence of arsenic and then allow it to cool. The surface acquires a silvery deposit, readily fooling the onlooker into believing the copper has turned to silver. Run away before they polish the 'silver' and the deposit rubs away.

Having convinced your punters you can turn base metal into gold or silver, persuade them to invest in your project – then abscond with the funds.

A USEFUL FRIEND

The Spanish alchemist Ramon Llull (*c.*1232–1315) is credited with several alchemical texts that he certainly did not write (he was one of the alchemical refuseniks) and – more fabulously – with being able to turn himself into a red cockerel. He is also said to have turned 22 tons of base metal into gold for King Edward III of England in order to finance a war against the Turks. It is reported that he put restrictions on the use of the gold, saying that the king must not use it to fight against other Christians – restrictions the king ignored, promptly launching an attack on France. Details of the legend undermine it immediately, not the least of which being that Edward III was still an infant when Llull died.

The legend of Ramon Llull suggests that the Battle of Crécy, fought by Edward III's troops against the French in 1346, was funded by misappropriated alchemical gold.

2
3
4
5
6
9
10
11
12
13
14
17
18
19
20
21
26
27
28
29
30
33
34
35
36
37
38
42
43
44
45
49
50
51
52
53
54
57
58
59
60
61
62

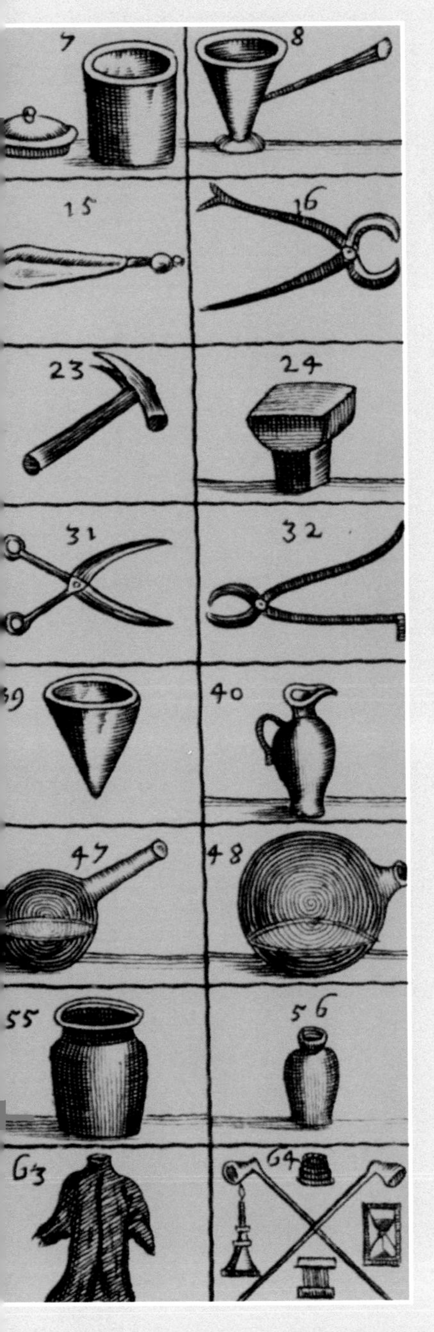

CHAPTER 3

GOLD AND THE GOLDEN AGE

'Many have said of Alchemy, that it is for the making of gold and silver. For me such is not the aim, but to consider only what virtue and power may lie in medicines.'

Paracelsus, 1493–1541

By the end of the Middle Ages, alchemy had a complex relationship with the Church but its standing as a science was unchallenged. Over the coming centuries it would attain its highest achievements and, finally, come to relinquish its place in the scientific world to the beginnings of modern chemistry.

Objects and instruments illustrated in plans for a 'portable laboratory' by alchemist Johann Becher (1635–82). Central to his plan was a portable furnace built of several components. Of his objects, only the bear's paw would seem alien to later chemists.

Renaissance: alchemy reborn

The Renaissance, a time of intellectual reinvigoration, is generally considered to have started in Italy in the 14th century and to have spread to northern Europe over the coming century. It was at its height in the 15th and 16th centuries and was marked by a resurgence of interest in learning, accompanied by growing confidence in human powers and abilities. This growing confidence was reinforced by accomplishments ranging from practical developments to the European discovery of the New World.

The invention of moveable type in the 1440s or early 1450s led to an explosion in

Michelangelo's iconic painting of the creation of Adam by God captures the spirit of optimism and potential that characterises the Renaissance.

NEW ALCHEMY FROM THE PAST

In the late 15th century, the great Italian humanist scholar and philosopher Marsilio Ficino (1433–99) translated into Latin for the first time the complete works of Plato and the *Corpus Hermeticum*. The *Corpus* consisted of fourteen texts, including alchemical dialogues in which Hermes Trismegistus apparently instructs a disciple. At the time of their translation, they were thought to be considerably older than they really were, perhaps predating Plato and coming from Ancient Egypt. Some recent scholars have suggested that parts could indeed have their origins in much earlier texts, but the *Hermetica* as presented were composed in Graeco-Roman Egypt and date from the 2nd and 3rd centuries AD.

Marsilio Ficino's accomplishment in translating the works of Plato had an immense impact on 15th-century European culture.

> *'This century, like a golden age, has restored to light the liberal arts, which were almost extinct: grammar, poetry, rhetoric, painting, sculpture, architecture, music . . . this century appears to have perfected astrology.'*
>
> Marsilio Ficino, 1492

literacy and the dissemination of written material, making the spread of learning easier and faster than it had ever been. Among the books printed were, inevitably, alchemical tracts and treatises. A 'golden age' of alchemy was soon underway.

Back to basics

The basic principles in alchemy at the start of its golden age were still grounded in theories from Aristotle and the Arab scholars about the nature of matter and how it occurs and changes naturally. By the 16th century, the received wisdom of the ancients was finally being challenged in many areas of practical science – medicine, anatomy and 'natural philosophy' (the umbrella term for the non-medical sciences). But alchemy remained immune to this critical examination for some considerable time – at least until the late 17th century. Alchemy shared basic ideas with geology that still held firm (the creation of metals within the earth over centuries or millennia). There was no evidence from observation or investigation to challenge their credibility, as there was in anatomy, biology and physics. The notion of transformation was based on reason, not just wishful thinking; unfortunately, the reasoning was wrong.

The temptation to engage the help of demons in academic and alchemical pursuits could be hard to resist, as the legendary figure of Faust discovered to his cost.

Angels and stones

Even though the quest was rooted in what seemed like sound science at the time, it was not without its superstitious aspects. The two – practical science and the supernatural – were not distinct and separate realms. The influence of the stars was a real and a serious consideration, so that some processes had to be carried out or materials collected at particular phases of the Moon or alignments of the planets. Astrology was considered to have the same scientific validity as using the right materials.

Alchemists sought out recipes and expert advice, not just from one another and printed works, but even from angels and spirits. Praying for inspiration or aid before a difficult endeavour was far from unusual at the time, and no sensible traveller would set out on a hazardous journey without first asking for God's favour. Even the chemist Robert Boyle mentioned contacting spirits for advice about making the philosophers' stone in the 17th century, and John Dee, known as a magician as much as a scientist,

GOD'S ON SIDE

The Italian scholar Giambattista della Porta (1535–1615) thought lessons in God's creation would help the alchemist in his practice:

'Both the vessel and the receiver must be considered according to the nature of the things to be distilled. For if they be of a flatulent nature and vaporous they will require large and low vessels and a more capacious receiver. . . . But if the things be hot and thin you must have vessels with a long and small neck. Things with a middle temper require vessels of a middle size. All which the industrious artificer may easily learn by the imitation of nature, who have given angry and furious creatures as the lion and bear thick bodies and short necks; to show that flatulent humours would pass out of the vessels of a large bulk and the thicker part settle to the bottom; but then the stag, the ostrich, the camelopard [giraffe], gentle creatures and of thin spirits, have slender bodies and long necks; to show that thin and subtile spirits must be drawn through a much longer and narrower passage and be elevated to purify them.'

is reported to have consulted angels with the help of the 'scryer' Edward Kelly and his 'shew stones', which revealed the angels.

It was a small step from enlisting the help of angels to taking the assistance of demons, though, and the polymath and scientist Athanasius Kircher (1602–80) warned against the temptation of invoking demonic aid when the alchemist's legitimate attempts were frustrated.

Getting down to business

There was dispute about the initial ingredients for making the philosophers' stone, but there was general agreement about the process itself. By the 17th century, most alchemists agreed that organic materials such as blood or urine were not suitable. Metals or minerals were the usual starting point. No doubt alchemists tried all options at one time or another.

The basic method was relatively simple, if laborious, though the details were often omitted from written records. The chosen matter was put into a flask with a long neck and 'hermetically sealed' by melting the glass of the neck and fusing it so that it was airtight. The flask, called an 'egg' on account of its role in the generation of the stone, was then heated at a constant temperature for a long time – months, in fact. This was no mean feat in the days of simple ovens and furnaces and before the invention of the thermometer. It was also not safe, as sealed glass containers were prone to explode on heating, causing at best a lot of mess and at worst considerable injury, as Giambattista della Porta remarked: 'when the heat shall have been raised up to the flatulent matter and that find itself straitened in the narrow cavities, it will seek some other vent and so tear the vessel to pieces (which will flie about with a great bounce and crack, not without endamaging the standersby).'

After a long interval, the contents of the egg would turn black, indicating successful completion of the first stage of transformation. Then the proto-stone

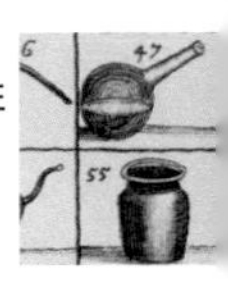

would display myriad colours, a stage sometimes called the peacock tail. After that, it would turn white. The alchemist could break the seal at this point and carry out some additional stages to fix an elixir for producing silver. Most continued to heat the egg until it darkened to yellow and eventually deep red – the final stage. The jubilant alchemist could then remove it from the sealed flask and mix it with gold and philosophical mercury to create the final stone, a dense, brittle, deep red material that could suffuse other matter.

This basic grade of philosophers' stone was thought capable of transmuting about ten times its own weight of base metal. The alchemist had to melt the base metal, or if using mercury heat it nearly to boiling point, then throw in a bit of the stone weighing about a tenth of the weight of the metal and continue heating. The contents of the crucible should then turn to molten gold. The diligent alchemist could refine and concentrate it by heating with mercury and go through the colour-changing sequence again to increase its power tenfold.

The process could be repeated endlessly. The most powerful stone was reputedly found by John Dee in a tomb and could transmute 272,330 times its own weight of base metal. Even granted the implausibility of the stone's existence, this is extravagant. It means that if the alchemist used just a hundredth of a gram he would need to melt 2.7kg (6lb) of base metal, which would itself be a prodigious feat.

A 19th-century representation of John Dee performing an experiment for Queen Elizabeth I.

Alchemy was a hot and bothersome business. This depiction of an alchemical laboratory from 1532 shows much of the alchemist's equipment clearly, some of it unchanged since the Arab Middle Ages.

There were various explanations for how the transmutation effected by the stone might work and they were all natural in the sense that they were based on a physical or chemical model – they did not require the operation of magic or spirits. Occasionally, detractors might claim that magic was involved, but the practitioners were quite clear that they were harnessing natural powers of transformation, just like the transformation of grape juice into wine, or unrisen dough into risen dough ready for baking. It was a difficult process to get right, but it was not, in the alchemists' view, mysterious or supernatural.

Alchemy and medicine

While many alchemists were occupied with the search for a successful transformative agent, others were more concerned with the search for medicines.

TRAPPED IN TIME

An alchemists' laboratory in Prague (now in the Czech Republic) that had been used by alchemists in the service of Rudolph II was sealed up in 1697 when the French invaded and was not rediscovered until 2002. A flood brought down a wall in the oldest building in the Jewish quarter, revealing steps to the underground laboratory, still with its equipment, a sealed bottle of elixir and some recipes. The underground space dates from the 10th century. Analysis of the elixir suggests that 77 different herbs and spices were included in its preparation.

The medical chemist and iatrochemistry

Swiss physician Theophrastus Bombastus von Hohenheim, better known as Paracelsus (see box, opposite), took alchemy in a new direction. He expanded it to an entire worldview, in which everything that happens does so through the medium of chemical transformations. He saw God as a Master Chymist and saw all physical and biological activity as ultimately chymical in nature, including the formation of minerals, the growth of plants, the processes of reproduction and digestion and the weather. Even the final judgement day would, he thought, be a chemical extravaganza.

Chemistry and the body

Paracelsus had no interest in the transmutation of metals and even criticized those alchemists who sought to achieve it, but looked to the chemistry the alchemists used as a medical resource.

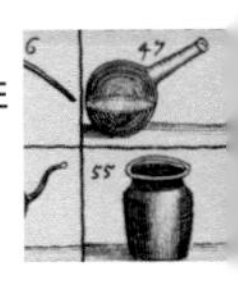

PARACELSUS (1493–1541)

Provocative and innovative, Paracelsus had a huge impact on 16th-century science.

Paracelsus was born in Switzerland and learned medicine from his father; he then travelled throughout Europe, rarely settling for long. As he travelled, he collected the wisdom and lore of everyone involved in medical practices, including barber-surgeons, midwives, gypsies and fortune-tellers – people usually disdained by traditional physicians. He built an impressive repertoire of chemical remedies, many produced from minerals rather than the more conventional plants, and many of which were toxic in larger doses.

He dismissed the medicine of his contemporaries, openly despising the teachings of Galen and Ibn Sina. Appointed professor of medicine in Basel, Switzerland, he began by publicly burning the books of these revered authorities and compounded the affront by lecturing not in Latin but in his native German.

Despite the highly practical nature of his chemical experimentation, Paracelsus also took on board a lot of mystical and astrological pseudo-knowledge. The mix of science and mysticism made his published work difficult to understand and his insistence on astral influences on chemistry alienated yet more of his contemporaries.

Paracelsus was a difficult individual – apparently arrogant, argumentative, openly aggressive and rude to those who opposed his new ideas. Yet he had immense influence, and his notion of the body as a chemical system would eventually come to be the dominant paradigm, one that persists to the current day.

He invented the term 'spagyria' to denote the process he favoured for refining and purifying matter through separating and recombining the components. The process involved using heat to separate out the three fundamental elements of matter, which he took to be mercury, sulphur and salt. Next, he aimed to recombine them, leaving out the impurities that he had separated from them. He believed that by this method even poisons could be purified and used as medicines, since the toxicity was all in the impurities. The principle of refining to remove impurities he called Scheidung. This model had the benefit of aligning the chymist's work with the work of God with souls, elevating his worth.

Paracelsus's insistence on the efficacy of chemical remedies produced a rift in medicine and initiated the field of iatrochemistry. The traditional practitioners, who relied entirely on remedies based on living material, rejected his inorganic approach, but a growing band of new medics adopted his views. Hostility

and competition between the two groups was open and often pronounced. Traditionalists were (unsurprisingly) not convinced by the Paracelsians' extensive use of toxins such as mercury and antimony, which they saw as dangerous. Only when King Louis XIV of France was successfully cured of an illness by an antimony emetic did the Paris medical school approve the use of antimony, and even then only because they were left with no choice.

Balance, humoral and chemical

Paracelsus was ahead of his time in many ways. He considered health to depend on balance and harmony in the human microcosm and the macrocosm of Nature. This was not a spiritual harmony, or the Galenist balance of humours (which he considered absurd), but an actual balance of minerals and other chemical substances. This balance could be redressed when upset by the administration of medicines containing the chemicals that were lacking.

The Galenists treated illness in generic ways according to a diagnosis of humoral imbalance. It might include, for example, applying hot baths to increase the heat and moisture of the body, bleeding to reduce the amount of blood, or emetics to remove bile. Each treatment could be used for many different types of illness. The approach taken by Paracelsus was more nuanced, targeting specific illnesses. He advocated the use of poisonous substances in medicine, including mercury and arsenic, therefore it was quite easy for things to go wrong.

Paracelsus often prepared medicines from plants and minerals, but his chemical preparation of plant material aimed to produce a concentrated form that need be given in only small doses. He explored the herbal preparations advocated in folk medicine, reporting, 'I have not been ashamed to learn from tramps, butchers and barbers.' Despite his enthusiasm for empirical observation and experimentation, he was quite willing to believe in faeries, gnomes and spirits.

An Italian drug jar to hold mercury, used to treat syphilis and (less successfully) plague.

He scored notable successes, among them the use of mercury to treat the relatively new disease of syphilis. Syphilis is caused by a spirochaete bacterium,

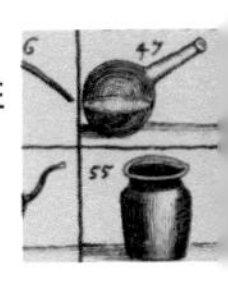

TRYING IT OUT – A MEDICINE FROM ANTIMONY

The chymists engaged in medical formulations often left very precise instructions. Though they were sometimes cryptic, in the manner of the alchemists, they did not leave bits out if their aim was to produce a useful medicine. The historian of chemistry Lawrence Principe successfully followed a recipe in *The Triumphal Chariot of Antimony*, published in 1604 by someone calling himself Basil Valentine (though this was probably a pseudonym). The recipe explains how to make a medicine from antimony by removing all its toxicity. Antimony is a metalloid and was of great interest to alchemists and chymists, perhaps on account of its mix of metal and non-metal properties.

The recipe begins with stibnite, an ore that contains antimony, and progresses by producing yellow 'glass of antimony' and then a red liquid that is further processed to give the medicinal 'sulphur of antimony'. Principe followed Valentine's instructions to grind up and roast the ore until it turns grey, then melt the 'ash' in a crucible, pouring it out to produce yellow glass. It didn't work. The grey ash was produced – the antimony is calcinated, forming an oxide – but the next stage produced only a grey lump. After numerous attempts, Principe followed a more precise detail of Valentine's instructions, and used stibnite from Eastern Europe (Valentine specified Hungary). This time it worked. Analysis showed that the East European stibnite contained a small amount of quartz. When he tried again with his original stibnite, but adding a pinch of powdered quartz, Principe succeeded in producing yellow glass. Valentine's recipe worked, but depended on an impurity.

The next step was to powder the glass and extract it with vinegar to produce a red solution. Again, Principe found that it didn't work unless he used the East European ore, when a slight pink tinge appeared. Further analysis revealed traces of iron in the original ore. And further reference to Valentine showed that he said he prepared his roasted antimony and then his solution with iron tools. As he stirred them, iron transferred from his tools to his mixture, depositing enough to make this solution red. Adding iron, again, gave Principe the right result.

The instructions then said to boil down the red solution to a sticky residue and make a solution with alcohol. Valentine's text assured him that the toxicity would remain in the residue and the medicine would be sweet and harmless. Indeed this was true. The antimony compounds were not soluble in alcohol, and remained in the part that was discarded. The final solution was of iron acetate, which has a sweet taste and is non-toxic. Valentine's 'antimony' medicine could be prepared following his instructions to the letter – but it contained no antimony.

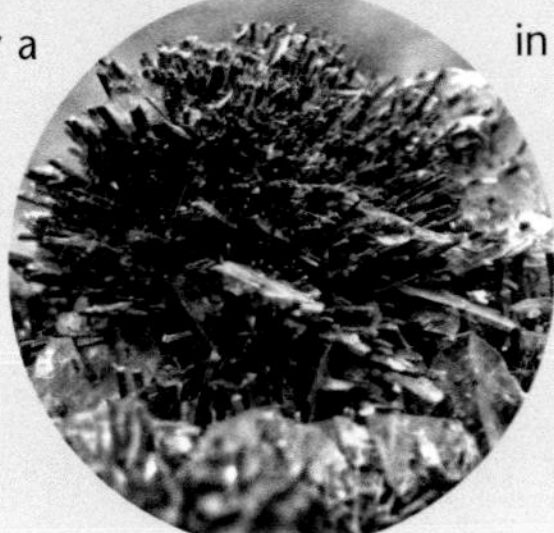

Stibnite, a mineral comprising antimony and sulphur (Sb_2S_3).

Treponema pallidum, which is poisoned by mercury. Although the treatment didn't work in a way that Paracelsus understood, it was effective – but dangerous for patients. Other innovations included advocating wounds be kept clean to prevent infection (a radical approach at the time) and his belief that illness has external causes rather than being the result of humoral imbalance.

Transcendent medicine

Most modern physicians would set the limit of medicine at preserving life. But the more ambitious alchemists sometimes made attempts at resurrection or the generation of life from inert matter.

Transforming lifeless to living

In the 16th and 17th centuries, spontaneous generation was still universally accepted: people believed that maggots, worms, flies and sometimes even snakes and crocodiles generated spontaneously from mud, rotting food and so on. It must have seemed conceivable, then, that an alchemist could put together the right combination of exotic materials and conditions to produce more advanced life, or to bring back life or the semblance of life from the dead matter of a previously living plant or animal. Even so, the Church would probably not have looked kindly on such a project.

A homunculus (a miniature living human) was attempted by some alchemists. Paracelsus (or one of his followers) describes the process of making a homunculus in *De natura rerum* (*On the Nature of Things*) in 1537. One method is to collect some human semen and seal it in a fairly large glass flask (you will need room for the homunculus to grow), then keep it in an incubator of horse manure for 40 days until it starts to coagulate. Thereafter, feed it human blood for 40 weeks, still keeping it warm. There's disagreement about whether the homunculus will arise already gifted in many arts or whether, as Paracelsus advises, it needs to be educated. Letting it out of the jar for too long can kill it, so homunculi are generally shown in their jars.

The lion eating a snake in this alchemist's laboratory is not an exotic, hungry pet but part of the obscure iconography of allegorical alchemical imagery.

Secrecy, truth and fraud

Alchemical practices such as making homunculi, transmuting metals into gold

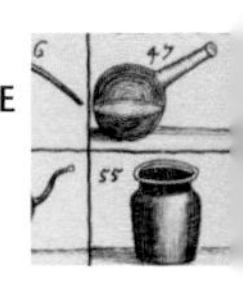

and producing healthful elixirs were highly controversial. They provided plenty of potential for fraudulent alchemists to scam the gullible, and for opponents to denounce alchemical practices as heretical or illegal. The secrecy that still surrounded these practices was equally effective at disguising a supposed truth, as a protective mechanism or a means of concealing fraud.

A conspiracy of secrecy

Given the atmosphere of mixed hostility and curiosity, it is no surprise that alchemists continued to record their work in esoteric and obscure ways, using a host of symbols and special terminology, codes and metaphors. This not only served to protect them, but also contributed to the air of mystery and exclusivity that attached to alchemy. Obscure language has often been used as a tool of social exclusion and to signal membership of an intellectual elite, and alchemists were not immune to its allure.

Illustrations such as this are impossible to decode without knowledge of the iconography of alchemy.

False trails

From the time of the Arab alchemists, there had been a tradition of splitting the information into discrete parts to be kept in different places. This way, only those truly dedicated to the pursuit would track down all the parts and put them back together; casual discoverers of part of the secret would not get far.

The metals and heavenly bodies shared symbols that were used widely in the alchemical texts. By substituting the personification of the appropriate astronomical body for a metal, it was easy to construct allegorical narratives or images.

With the development of printing in the West, the use of allegory and metaphor to disguise aspects of alchemical knowledge took a new turn. Printed books made increasing use of illustrations, and some presented allegorical images of alchemical processes and ideas. Hermaphrodites, and couples engaged in coitus, were common components of these alchemical images – a natural extension of the idea of generating new material by mingling contrasting types of original. Would it be too cynical to suggest that there might have been some additional thrill for the alchemist in being licensed to look at images of licentious behaviour? Alchemy porn – or at least erotica – was maybe a perk of the job.

Cryptic gold-making

Allegorical alchemical narratives, often accompanied by cryptic illustrations, are difficult to interpret today. It would be easy to assume it's all a lot of mumbo-jumbo, either deliberately obfuscating procedures the authors knew to be nonsense or presenting something rather woolly and ill-defined. But this seems not to be the case.

Principe – who made the antimony-based medicine – also tried to decipher the allegorical text and imagery in Valentine's book *Of the Great Stone*, divided into twelve 'keys' to alchemical success. Each key is a stage in the process of preparing the philosophers' stone. Each one is presented as an allegorical narrative, in which the names of things that stand for gold, silver and other chemicals change frequently, and is accompanied by an allegorical woodcut. Principe has unravelled obscure instructions such as, 'When [the ravenous grey wolf] has devoured the king, then make a great fire and throw the wolf into it so that he burns up entirely,' turning them into practical laboratory instructions. In this case, it means that gold (the king) is thrown into molten stibnite (the wolf), which will dissolve (devour) it, and it should then be heated. It is true that gold melts readily in molten antimony and, as the instructions go on to explain cryptically, that the gold can easily be recovered from the antimony (the king will be redeemed when the wolf is burned). The allegorical presentation succeeds in concealing chemical knowledge from the ignorant, but revealing knowledge to the initiated. The procedure is rooted in careful practice as some of the techniques, such as the sublimation of gold, are very difficult to achieve, even in a well-equipped modern laboratory.

Of course, the alchemists did not succeed in producing the philosophers' stone (we assume). So somewhere along the line the instructions given by Valentine must change from what he has achieved

himself to practices others claimed to have achieved or those believed in theory to be possible. But plenty of people believed transmutation to be possible. It must have been like an urban myth: 'My friend knows someone who made the philosophers' stone . . . ' Many success stories survive and museums around Europe have medals and coins cast from gold supposedly produced by alchemy – with inscriptions attesting to their provenance.

Fraud and failure

Valentine's instructions could have landed some alchemists in a sorry mess. While fraudsters abounded, many alchemists genuinely believed they could produce gold by following his instructions, if only they could secure the necessary materials, premises and tools. They entered into contracts with wealthy and powerful men only to find that the recipe did not work and they were in breach of contract. In Germany, particularly, unfortunate aspirant alchemists were generally executed. Even if the alchemist had honest intentions, failure was classed as fraud for which the penalty was death.

A woodcut introducing Key 9 of Basil Valentine's Of the Great Stone.

A TINSEL GOWN AND A GOLDEN GALLOWS

Public acceptance of the possibilities of transmutation and of making the philosophers' stone continued into the 18th century. Domenico Caetano (1667–1709), an Italian peasant who had learned both metal-working and conjuring, passed himself off as a German count and rose to high office. He claimed to have discovered a manuscript explaining how to prepare the philosophers' stone, and began demonstrating transmutations. These were evidently so convincing that in Brussels he secured a sizeable fortune in advance for transmutation. Eventually Caetano was caught and imprisoned for six years, but managed to escape. After another series of rash promises – this time to make a large quantity of the philosophers' stone for Frederick I – and attempting to escape, he was executed. Caetano was hanged wearing a cloak of gold tinsel from a gallows itself covered with gold. Gold medals were struck to commemorate his execution.

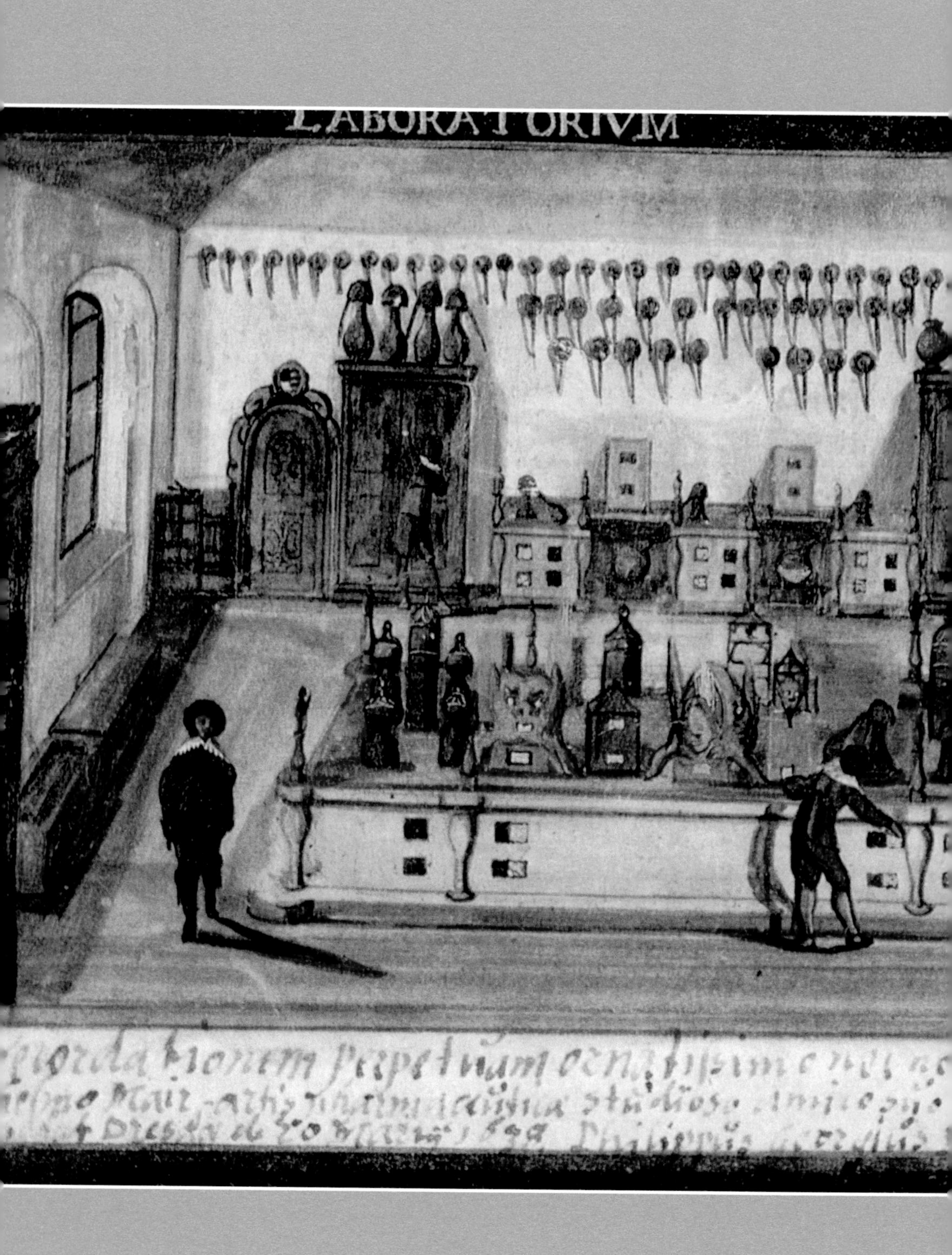
LABORATORIUM

CHAPTER 4

FROM ALCHEMY TO CHEMISTRY

'Although nature begins with the cause, and with experiment, we must do it inversely, we must discover the cause with experiments.'

Leonardo da Vinci

In the 17th and 18th centuries, chemistry began to branch off from alchemy. While alchemy was rooted in philosophy and theory, chemistry was distinguished by its foundations in the physical world and in experimentation. Chemistry emerged as a science at last.

Inside a chemist's laboratory – with monsters – in 1638.

The scientific method

Aristotle was the first theorist to suggest that we should rely on the evidence of our senses in understanding how the world around us works. Unfortunately, his more specific conclusions were treated with such reverence that his trust in the empirical method was largely forgotten. Errors in Aristotle's science were not overthrown by later observation and experimentation. Instead, evidence contrary to established authorities, including Aristotle, was distrusted, misconstrued and misinterpreted. It took a major paradigm shift to free the sciences from the shackles of the past.

Scientific revolution

During the early Renaissance, science remained rooted in classical authorities, principally Hippocrates, Empedocles, Aristotle, Galen and Ptolemy. But in the 16th century this began to change. Anatomists from Andreas Vesalius onwards explored the human body through dissection. They contested Galen's account when they found contradictions between the texts and the evidence of their eyes and experience. Nikolai Copernicus and Johannes Kepler challenged Ptolemy's model of a universe revolving around a central Earth. In 1543, Copernicus showed that the Sun lies at the centre of the solar system and, in 1609, Kepler demonstrated that the orbits of the planets around it, including Earth, are elliptical. A supernova in 1572 and a comet in 1577 showed that the heavens are not, after all, eternally unchanging. Around 1600, the invention of both the microscope and the telescope changed forever the way we see the world and the universe. The microscope repopulated the world with an infinite number of tiny, unimagined beings, and the telescope showed details of planets that had previously been seen as mere spots of light. Galileo Galilei (1564–1642) and Isaac Newton (1642–1726) found mathematical laws that explained and predicted the behaviour of the physical world.

The appearance of a new star in 1572, observed here by astronomer Tycho Brahe, challenged the belief that the heavens are unchanging and paved the way for modern astronomy. The 'star', which subsequently disappeared again, was a supernova (SN1572).

> **INDUCTIVE AND DEDUCTIVE METHODS**
>
> There are essentially two different ways to approach the search for knowledge and meaning. We can begin with careful observation and try to tease out the general rules which lie behind the phenomena. Or we can start with philosophy or theory and try to interpret observed phenomena in the light of that. The first is inductive reasoning and the second deductive reasoning. Until the 16th century, most scientific enquiry took the second path; people had a pretty good idea – they thought – of how the world worked, and then operated within that model to try to explain observed phenomena. With the rise of the scientific method, there was a shift towards using an inductive method – using observation and active experimentation to collect data from which rules about the world could be worked out.

These discoveries unpicked certainties rooted in the authority of the ancients and even contradicted the Bible. It was a time of intellectual uncertainty and crisis as well as excitement, a scientific revolution that overturned not just the established models but an entire way of thinking.

It took rather longer in chemistry for new models to replace the old, and alchemy continued to flourish in tandem with the emerging modern chemistry in a synthesis now called chymistry. The scientific revolution is generally dated around 1550–1700, but the revolution in chemistry began towards the end of this period and was most dramatic in the 18th century. While chemistry remained tied to the philosophical roots of alchemy, real progress was stifled; new discoveries were interpreted in the light of an incorrect model. It took a profound shift in outlook – the confidence to allow a new model to emerge from experimental results and observations – for this to change.

The science of Bacon

In 1620, the English philosopher Sir Francis Bacon (1561–1626) laid the foundations of what has become known as the scientific method. He suggested that scientists – or 'natural philosophers', as they were then called – should adopt a rigorous process of observation and experimentation to test philosophical ideas. Their approach should be critical and investigative; they should not accept a widely held belief as true without interrogating it. His method advocated starting with doubt, whereas the prevailing deductive method started with certainty. The book in which he proposed this approach was called the *Novum Organum* after Aristotle's *Organon*, which encouraged scientists to work from observation to general laws. He went further than Aristotle, though, in also advocating active experimentation.

The central part of Bacon's recommendation is that science should start with a hypothesis that is tested by regulated and repeated investigation. The conclusions the scientist draws should not go beyond those which the evidence directly supports. For instance, to test the idea that cold, wet conditions cause illness, healthy people should be subjected to such conditions and

Sir Francis Bacon, often considered the originator of the scientific method.

their health checked afterwards. If they were found to fall ill, this would not automatically support the theory of humours, but only show that – for some as yet undetermined reason – cold and wet conditions cause illness. (Obviously this is rather unethical, especially when there were relatively few effective treatments for illness. Bacon's ideas were not always something that could be put into practice.)

Change was already afoot in Europe, with the received knowledge of millennia being questioned at last. What Bacon suggested was a framework in which these challenges to authority and discoveries could be managed, and in which philosophers could be more confident that they were building towards truth, not simply another set of mistaken premises.

Learning and societies

Bacon called for an institution to be founded that would promote and regulate knowledge according to his method, providing a sort of quality control mechanism for new scientific knowledge. It was many years later, in 1660, that the Royal Society for Improving Natural Knowledge was founded in London as an 'invisible college' with the motto '*Nullius in verba*', taken to mean 'take nobody's word for it'. Later it became just the Royal Society (and still exists as such) – the first of the learned societies that would spring up all over Europe to regulate knowledge and professional behaviour in specific areas of expertise. The Royal Society would take a lead in distinguishing between alchemy and chemistry.

Alchemy and the scientific method

Alchemy, quite clearly, does not begin from observed phenomena and proceed to generalized laws. It begins with the belief that matter is made up of a few fundamental elements and can be reconfigured by the addition or removal of elements or properties to transform it. All that follows is based on this belief, and failure to effect transformation was taken as incompetence on the part of the alchemist or using the wrong method, rather than a prompt to question the underlying model.

NEWTON'S REGULAE

Isaac Newton set out in his *Principia Mathematica* (1687) four rules that he believed should govern the scientist's treatment of knowledge derived from observation.

- No more causes of natural things should be admitted than are both true and sufficient to explain their phenomena.
- Therefore, the causes assigned to natural effects of the same kind must be, so far as possible, the same.
- Those qualities of bodies that cannot be intended and remitted and that belong to all bodies on which experiments can be made should be taken as qualities of all bodies universally.
- In experimental philosophy, propositions gathered from phenomena by induction should be considered either exactly or very nearly true notwithstanding any contrary hypotheses, until yet other phenomena make such propositions either more exact or liable to exceptions.

The scientific method drove a wedge between chemistry and alchemy. Although alchemical practices did not cease immediately, they became distanced from science. Alchemy became more esoteric and chemistry more rooted in the empirical. Separation was inevitable.

An amicable divorce

Some of the great scientists of the time took a serious professional interest in alchemy and saw no contradiction in pursuing it alongside their non-speculative science.

Alchemical chemists

Along with Newton, the chemist Robert Boyle (1627–91) and the biologist and chemist Jan Baptist van Helmont (see page 76) stand out as men known primarily for their achievements in conventional natural sciences, but who were also enthusiastic alchemists. Indeed, Newton expended more time and energy on alchemy than on physics and mathematics, writing in total more than a million words on the subject.

One character who did much to encourage both Boyle and Newton was George Starkey (originally Stirk). Born in Bermuda, he studied at Harvard then

The newly formed Royal Society met first in offices in Crane Court, off London's Fleet Street.

moved to London, England, where chemical supplies and colleagues were easier to come by. He was a chymist, making and selling iatrochemical medicines (that is, treatments with origins in alchemy), perfumes and chemical furnaces to fund his alchemical experiments.

Starkey presented himself as a conduit to an alchemist he called Eirenaeus Philalethes, and started producing documents under this name. Starkey claimed that Philalethes could make the philosophers' stone and had given him samples. In his private correspondence Starkey described his chymical practices openly, but in his writing as Philalethes he was as obscurely allegorical as any alchemist. He was a huge influence on Boyle, the leading chemist of his day, writing letters to him that Boyle passed on to others, including Newton. In 1665, at the age of 37, Starkey died of plague – a disease against which his iatrochemical remedies proved ineffective.

Starkey was not alone in his influence. In his notes, Boyle recorded several examples of demonstrations of transmutation, including one around 1680 at which he was even invited to cast fragments of the red philosophers' stone into molten lead himself. He declined, afraid that his hand would shake and he would drop the precious substance into the fire, but he witnessed the transformation which apparently occurred when the alchemist showed how it was done. Boyle had the metal tested and found it to be pure gold. He was so convinced by this and further demonstrations that in 1689 he testified before Parliament to the reality of transmutation in support of a bid to have the 1404 ban on the practice repealed. The bid was successful, and the transmutation of base metals into gold was legalized in England in 1689.

However, the tide was turning. After Newton's death in 1727, the Royal Society deemed his alchemical works 'not fit to be printed', so they disappeared into an archive where they languished for nearly 300 years. Nor was Newton keen to publicize his work. He wrote to Boyle of the need to maintain 'high silence' about their art and discoveries.

BOYLE TRICKED

Boyle was not always lucky in his dealings with alchemists. In 1677, a Frenchman named Georges Pierre des Clozets visited him in London and introduced him to an international society of alchemists called 'Asterism'. After some correspondence, and sending gifts to someone represented as the 'patriarch of Antioch', Boyle was invited to join the group. Pierre des Clozets was to act as his representative at a meeting in a castle near Nice, France. Among the wonders promised by the society was a Chinese alchemist who was said to have a homunculus (see page 64) which lived in a glass jar. But Boyle's membership came to nothing; according to Pierre des Clozets, the castle was blown up by a bomb and many of the members killed. Boyle then discovered that des Clozets had not even travelled to Nice, but failed to get his own valuable papers returned. Pierre des Clozets died in 1680.

BOYLE'S SCIENTIFIC WISH-LIST

Boyle produced a list of 24 things he wished to see invented or achieved, most of which have come to pass at least to some degree:

- The Prolongation of Life.
- The Recovery of Youth, or at least some of the Marks of it, as new Teeth, new Hair colour'd as in youth.
- The Art of Flying.
- The Art of Continuing long under water, and exercising functions freely there.
- The Cure of Wounds at a Distance.
- The Cure of Diseases at a distance or at least by Transplantation.
- The Attaining Gigantick Dimensions.
- The Emulating of Fish without Engines by Custome and Education only.
- The Acceleration of the Production of things out of Seed.
- The Transmutation of Metalls.
- The makeing of Glass Malleable.
- The Transmutation of Species in Mineralls, Animals, and Vegetables.
- The Liquid Alkaest and Other dissolving Menstruums. (This means a universal solvent.)
- The making of Parabolicall and Hyperbolicall Glasses.
- The making Armor light and extremely hard.
- The practicable and certain way of finding Longitudes.
- The use of Pendulums at Sea and in Journeys, and the Application of it to watches.
- Potent Druggs to alter or Exalt Imagination, Waking, Memory, and other functions, and appease pain, procure innocent sleep, harmless dreams, etc.
- A Ship to saile with All Winds, and A Ship not to be Sunk.
- Freedom from Necessity of much Sleeping exemplify'd by the Operations of Tea and what happens in Mad-Men.
- Pleasing Dreams and physicall Exercises exemplify'd by the Egyptian Electuary and by the Fungus mentioned by the French Author.
- Great Strength and Agility of Body exemplify'd by that of Frantick Epileptick and Hystericall persons.
- A perpetuall Light.
- Varnishes perfumable by Rubbing.

It would be 200 years before the realization of Boyle's dream of humans attaining 'the Art of Flying'.

For the faithful, Holy Communion enacts a transformation even more wondrous than those promised by alchemy.

Transformations all round

The point at which alchemy and chemistry began to go their separate ways is marked more by a shift in approach than by changing beliefs about what was possible. In the 17th century, mysterious transformation was widely accepted. The Catholic Church held that the bread and wine of the sacrament were transformed literally into the body and blood of Christ: it was a daily miracle which every Catholic accepted without a second thought. Metals in the ground seemed to transform from one type to another, with seams of gold appearing in rock. Rotting food and other matter seemed to transform into living maggots, worms, flies and even scorpions and mice. And, of course, food, water, sunlight and earth are daily transformed into the bodies of plants and animals. It would have appeared that the evidence of the senses supported strange transformations.

The significant change came with applying the scientific method and inductive reasoning to investigating these transformations. This great paradigm shift would eventually be the undoing of alchemy: nothing found in the laboratory supported the notion that base metals could be turned into gold, or upheld the philosophical framework of alchemy.

Van Helmont's tree

One experiment perfectly straddles the boundary between alchemy and chemistry, a marker of both the rise of the scientific method and the continuation of old theories. It was carried out by the Flemish scientist Jan Baptist van Helmont (1580–1644), a believer in alchemy. Van Helmont was convinced that the original matter is water, just as Thales had claimed 2,000 years previously. He proposed that as a plant grows it converts water into the bark, leaves, roots, seeds and so on of its body, and designed an experiment

Van Helmont's tree was a pivotal point in the history of the scientific method.

'By this apparatus I have learned that all things vegetable arise directly and in a material sense from the element of water alone. I took an earthen pot and in it placed 200 pounds of earth which had been dried out in an oven. This I moistened with rain water, and in it planted a shoot of willow which weighed five pounds. When five years had passed the tree which grew from it weighed 169 pounds and about three ounces. The earthen pot was wetted whenever it was necessary with rain or distilled water only. It was very large, and was sunk in the ground, and had a tin plated iron lid with many holes punched in it, which covered the edge of the pot to keep air-borne dust from mixing with the earth. I did not keep track of the weight of the leaves which fell in each of the four autumns. Finally, I dried out the earth in the pot once more, and found the same 200 pounds, less about 2 ounces. Thus, 164 pounds of wood, bark, and roots had arisen from water alone.'

Jan Baptist van Helmont, *Ortus Medicinae* (1648, published posthumously)

to test his theory. (The prevailing theory was that plants take material only from the soil to build their structures.) It was the first time the scientific method had been applied to biology, or the chemistry of a living organism – at least, it was the first time the results had been published. Leonardo da Vinci (1452–1519) had carried out the same experiment with pumpkins, but recorded the results in his private notebooks only.

Applying the scientific method to the question, van Helmont set out to test his theory that plants grow only from water. This was courageous, as in 1634 he had been arrested and questioned by the Spanish Inquisition for studying plants and other natural phenomena. First, he weighed a willow sapling, then he weighed a large pot of dried soil. He planted the sapling in the pot, covering it so that nothing could blow into the pot, and watered it. He looked after the tree for five years. At the end of that time, he carefully emptied the pot, removing the soil from the tree's roots, and weighed the soil again. And he weighed the tree. The tree had gained weight – 74.3kg (164lbs) but the soil had lost only 60g (2oz). He concluded that the tree had not grown from soil but from the water he had supplied.

Van Helmont's conclusion was wrong: water is important to plants, but they take their chemical building blocks mostly from the gases of the air, with small quantities of essential nutrients from the soil. Ironically, van Helmont's other claim to fame was the discovery of carbon dioxide (see page 100), but he didn't suspect the air of

Jan Baptist van Helmont.

supplying anything his tree might need. While wrong in the detail, van Helmont was right in concluding that plants break down and reconfigure the material they take in. But discovering the method of that transformation lay far in the future.

Chemistry emerging

If there is one defining moment that marks the emergence of modern chemistry, it is the publication of Robert Boyle's book *The Sceptical Chymist* in 1661. Boyle did for chemistry what Copernicus did for astronomy and Vesalius for anatomy – he lifted it out of the morass of accepted wisdom and proposed that things might not be exactly as everyone believed.

BOYLE'S PROPOSITIONS

Proposition I.
It seems not absurd to conceive that at the first production of mixt bodies, the universal matter whereof they among other parts of the universe consisted, was actually divided into little particles of several sizes and shapes variously moved.

Proposition II.
Neither is it impossible that of these minute particles divers of the smallest and neighboring ones were here and there associated into minute masses or clusters, and did by their coalitions constitute great store of such little primary concretions or masses as were not easily dissipable into such particles as composed them.

Proposition III.
I shall not peremptorily deny, that from most such mixt bodies as partake either of animals or vegetable nature, there may by the help of the fire be actually obtained a determinate number (whether three, four, or five, or fewer or more) of substances, worthy of differing denominations.

Proposition IV.
It may likewise be granted, that those distinct substances, which concretes generally either afford or are made up of, may without very much inconvenience be called the elements or principles of them.

Robert Boyle, *The Sceptical Chymist*, 1661

THE
SCEPTICAL CHYMIST:
OR
CHYMICO-PHYSICAL
Doubts & Paradoxes,
Touching the
SPAGYRIST'S PRINCIPLES
Commonly call'd
HYPOSTATICAL,
As they are wont to be Propos'd and
Defended by the Generality of
ALCHYMISTS.
Whereunto is præmis'd Part of another Difcourfe
relating to the fame Subject.

BY
The Honourable ROBERT BOYLE, Efq;

LONDON,
Printed by J. Cadwell for J. Crooke, and are to be
Sold at the Ship in St. Paul's Church-Yard.
MDCLXI.

Publication of Boyle's The Sceptical Chymist *was a pivotal point in the story of chemistry.*

Chemistry and scepticism

Boyle's book is presented as a dialogue between five friends on the structure of matter. In framing the debate in this way Boyle followed classical tradition, but his propositions (see box, left) overturned it. They stated that all matter is composed of tiny particles; that these are particles of fundamental elements; that the elements are not those set out by the Ancient Greeks, nor by Paracelsus; and that all the matter we see is composite – we don't see the elements in their pure form.

The insights in *The Sceptical Chymist* made modern chemistry possible. In particular, the dissociation of elements from earth, air, water and fire opened the door to the discovery of genuine chemical elements. Boyle defined elements without reference to any specific substances: 'I now mean by Elements, as those Chymists that speak plainest do by their Principles, certain Primitive and Simple, or perfectly unmingled bodies; which not being made of any other bodies, or of one another, are the Ingredients of which all those call'd perfectly mixt Bodies are immediately compounded, and into which they are ultimately resolved.'

This definition can stand without knowing what the elements are or how they combine. It also opens the door to two fundamental activities of chemistry: making chemical mixtures and compounds (chemical synthesis), and finding out the composition of chemical mixtures and compounds (analysis).

Elements old and new

By liberating chemists from the intellectual tyranny of the Greek legacy, Robert Boyle enabled the discovery of the real elements and took away the artificial constraints on how matter might be expected to react and interact. He didn't make it easy to identify the elements, however.

Of the substances that had been treated as elementary in some way, air, water, earth, salt or the non-substance fire are no longer considered to be elements. Mercury and sulphur, which the alchemists considered to be the physical components of metals, are recognized as elements. How then, did the new list of elements emerge?

Building a new list

Some elements have been known since prehistoric times, though not recognized as such. Gold, silver, lead, mercury, tin, copper, sulphur, antimony, arsenic, bismuth and zinc were known to the ancient

The iridescent colours of bismuth crystals belong to a very thin layer of oxide that forms over the surface in air.

civilizations (though not all were known to every civilization). Bismuth, although known, was often confused with lead and tin, and was only confirmed as a separate substance in 1753.

Between the ancients and the 18th century, just one new element was discovered – the first to be found through chemical experimentation. It was phosphorus, discovered by the German alchemist Hennig Brand in 1669, just eight years after the publication of *The Sceptical Chymist*. It was not identified as an element on its discovery.

Brand had been collecting urine (it's said he collected 1,500 gallons of it) which he used in his alchemical experiments. Urine had been used as an ingredient in alchemical preparations as far back as the Leiden and Stockholm Papyri, so there was nothing odd in this. What Brand did that no one else seems to have done before was to take a flask of urine and boil it down to a thick gooey mess, then leave this for a few months to become, one imagines, deeply unpleasant, then to heat it with sand and collect the fractions driven off, both gases and oils. The last fraction condensed to a white solid – phosphorus. At first, Brand thought he had discovered the philosophers' stone. White phosphorus emits a glow when in the presence of oxygen, the energy emitted as the element oxidizes, so it would have looked sufficiently spectacular to excite him.

Like most alchemists, Brand was secretive about his methods, even after he realized that phosphorus was not the philosophers' stone. He sold on his instructions for making it to a few people, including the German philosopher and mathematician Gottfried von Leibniz (famous for developing calculus independently of Isaac Newton). The method only became public after Brand's death, when someone sold his secret to the Academy of Sciences in Paris in 1737.

Brand's astonishing discovery of phosphorus, painted by Joseph Wright of Derby as The Alchymist in Search of the Philosophers' Stone *in 1771.*

Some living organisms use chemiluminescence – producing light from a chemical reaction.

More metals

After phosphorus, there was a gap of 60 years before the next element was found, but thereafter discoveries came thick and fast. In 1732, the Swedish chemist Georg Brandt found the first new metal element to be discovered. A professor of chemistry at Uppsala University, Brandt demonstrated that the blue colour often found in glass is produced by cobalt. Cobalt compounds had been unwittingly used in glazes to colour glass and ceramics for millennia. The chemists of the 18th century supposed the colour was produced by bismuth, which often occurs in the same ores as cobalt. Cobalt's identity as an element was not confirmed until 1753.

Brand was pleased by what he saw as the symmetry completed by his discovery: there were believed to be six true metals and now, with cobalt, there were also six part-

GOBLINS AND UNDERGROUND SPRITES

The name 'cobalt' comes from the German *kobold*, for 'goblin'. The ore containing cobalt was called kobold ore by miners on account of the toxic fumes produced when it was heated: it contains arsenic, which forms the volatile and dangerous arsenic oxide.

Nickel was also named after a mythological underground baddie. It was discovered in 1751 by the Swedish mining expert Baron Axel Cronstedt who named it *kupfernickel* as the ore looks like copper (*kupfer*), but miners proved unable to extract copper from it. Instead of recognizing there was no copper to extract, they blamed a sprite, or *nickel*, for thwarting their attempts. Cronstedt also discovered the mineral he named scheelite after Carl Scheele, and from which tungsten was later extracted.

metals (what we would consider metalloids). That symmetry would soon be wrecked.

Platinum was named as a new metal in 1748, nickel in 1751, bismuth in 1753 and magnesium in 1755. As with cobalt, some substances had been found and used previously, even though they had not been identified as discrete metals. Platinum was first described in 1557 by the Italian physician Julius Caesar Scaliger as a metal found mingled with South American gold. It is possible that some of the alchemists found platinum, too, as there are occasional reports of a metal about as heavy as gold that does not react with the usual acids. Platinum would fit this description, and could have been present in small, unnoticed quantities as an impurity in the gold the alchemists used.

Cattle will usually drink from almost any water source, so when Wicker's cows refused to drink from a particular waterhole, it was worth investigating.

Cows don't like it

The last of the metals to be found in this first batch of discoveries was magnesium. During a drought in 1618, an English farmer, Henry Wicker, found that his cows refused to drink from a water hole in Epsom, Surrey. Cows are not usually fussy, especially when thirsty, so Wicker investigated. The water from the hole not only tasted bitter, but had the unusual property of helping scratches and rashes to heal. The therapeutic benefits of 'Epsom salts' (magnesium sulphate) soon became widely known. In 1755, Joseph Black recognized that a component of Epsom salts (magnesium) was a new element. In 1808, the English chemist Humphry Davy (1778–1829) extracted magnesium for the first time by electrolysis, using a mix of magnesia (magnesium oxide) and mercury.

Thick and fast

The first gaseous elements were discovered in the mid-18th century, but then there was a gap of more than a century before the next was found. Progress was steadier with the metallic elements, yielding – after magnesium – in fairly rapid succession: barium (recognized in 1772, isolated in 1808), manganese (1774), molybdenum (1778/1781), tungsten (1781/1783),

tellurium (1782) and strontium (1787/1808). In 15 years, between 1789 and 1804, 14 new elements were found, including titanium and chromium. Sodium and potassium were both found in 1807, and calcium and boron in 1808.

Defining elements

Although retrospectively we see a rush of discoveries, at the time it was not clear which of these and other new chemicals were elements. The notion of an element was still somewhat fuzzy, despite Boyle's definition. Considerable clarity was brought to the issue by the great French chemist Antoine-Laurent Lavoisier (1743–94).

Lavoisier redefines chemistry

Lavoisier set out to clarify the nature of the elements and to lay the foundations of chemistry as a science in its own right; he is often considered the father of modern chemistry. He published *Traité élémentaire de chimie* (*Elementary treatise on chemistry*) in 1789 with the intention of promoting and explaining the 'chemical revolution' – the new ideas in chemistry that he and his contemporaries were pursuing, finally separating their work from that of the alchemists for good.

A new list

In the *Traité*, Lavoisier set out his definition of an element, or 'principle': a chemical substance that cannot be broken down by any method of analysis or decomposition. He readily acknowledged that the substances he listed might one day be further broken down, asserting only that this was not possible using current methods:

'Not that we are entitled to affirm, that these substances we consider as simple may not be compounded of two, or even of a greater number of principles; but, since these principles cannot be separated, or rather since we have not hitherto discovered the means of separating them, they act with regard to us as simple substances, and we ought never to suppose them compounded until experiment and observation have proved them to be so.'

Lavoisier listed 33 substances, 23 of which are still accepted as elements. Oddly, his list includes light and 'caloric', which he considered to be the mass-less substance of heat that caused other substances to expand in volume. These, alongside the gases oxygen, hydrogen and nitrogen, he considered 'elastic fluids'.

Lavoisier's other categories were non-metals, metals and 'earths'. The non-metals, which he defined as 'oxidizable and acidifiable non-metallic elements', were phosphorus, sulphur and carbon and also the roots of boric, hydrochloric and hydrofluoric acid (later identified as boron, chlorine and fluorine).

The 17 metals ('metallic, oxidizable, and capable of neutralizing an acid to form a salt') were silver, bismuth, cobalt, copper, tin, iron, manganese, mercury, molybdenum, nickel, gold, platinum, lead, tungsten and zinc, plus arsenic and antimony, the last two of which are no longer considered true metals but are elements.

The earths, or 'salt-forming earthy solids', are all now recognized as compounds (oxides of calcium, magnesium, barium, aluminium and silicon – the elements would be discovered over the coming century).

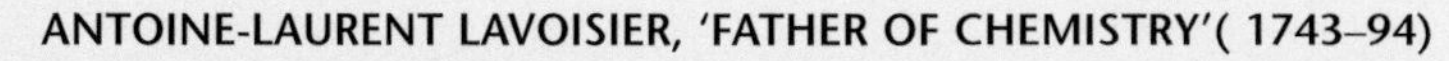

ANTOINE-LAURENT LAVOISIER, 'FATHER OF CHEMISTRY'(1743–94)

Born the son of a wealthy Parisian lawyer, Lavoisier trained in law himself but was really drawn to science. Throughout his life, chemistry was his chief passion though he also worked in taxation. He was a central figure of the Chemical Revolution of the 18th century.

Lavoisier's wife, Marie-Anne Pierrette Paulze, was only 13 when they married, but soon educated herself in science and learned English so that she could translate scientific papers for Lavoisier to read. She also learned illustration and engraving so that she could illustrate his papers and books.

In 1775, Lavoisier was appointed commissioner of the Royal Gunpowder and Saltpeter Administration and moved into the Paris Arsenal. There his well-equipped laboratory attracted young chemists from around Europe – and he considerably improved the manufacture of gunpowder. He made many significant discoveries, including the role of oxygen in respiration and combustion, and the chemical composition of water. His insistence on the conservation of matter as a guide to understanding chemical processes and reactions was crucial, and underpinned his meticulous attention to detail. He made careful measurements and recorded all his work.

Lavoisier debunked the theory of phlogiston (see page 94) and began the modern systematic naming of chemicals. In 1789, he published a seminal textbook, *Traité élémentaire de chimie*. But just five years later he was guillotined by the French Revolutionaries, accused of adulterating tobacco and of taking money from the national treasury to pay the enemies of France. Eighteen months later he was exonerated, the government admitting he had been falsely accused; his goods were returned to his widow.

Team Lavoisier: Antoine and his wife Marie-Anne worked together.

THE BLINKING HEAD

There is an apocryphal story that Lavoisier asked a friend to watch his head fall into the basket by the guillotine and promised he would continue to blink for as long as possible to see how long the head could survive decapitation. There is no evidence to support the story, and as Lavoisier and 27 others were tried, condemned and executed on the same day, the executions taking only 35 minutes, there would not have been time to conduct this final experiment.

Some of the substances do not fit into the category Lavoisier assigned to them (neither arsenic nor antimony is now considered a metal); and two are not even considered substances (light and heat). But it was a start. The most striking thing about Lavoisier's list was just how many elements he was prepared to allow. The ancient schemes had considered four or five to be sufficient building blocks from which to make the entire universe. Lavoisier's scheme was very different, allowing far more diversity. The rapid proliferation of elements soon after would become quite a headache for chemists, as we shall see later.

Lavoisier was just one of many people unjustly executed by guillotine during the French Revolution.

Trial and error

Lavoisier's principle that a substance must not be capable of further decomposition if it is to be considered an element is still sound, but the way it is conceived and tested has changed. The modern definition is that elements are composed entirely of atoms of the same type. So hydrogen is made of hydrogen atoms and zinc of zinc atoms, and so on. Lavoisier's 'element' hydrochloric acid doesn't count because it is made up of hydrogen atoms and chlorine atoms in combination: it is a compound. The problem for the 18th century chemists was that they had no way other than by experiment of determining which substances could be broken down into simpler components and which could not.

The next breakthrough would come at the very start of the next century. But before that, chemists had an entirely new chemical realm to discover and explore: that of the gases.

'It took them only an instant to cut off that head, and a hundred years may not produce another like it.'

Joseph-Louis Lagrange, commenting on the execution of Lavoisier, 1794

CHAPTER 5

AIRY NOTHINGS

'The importance of the end in view prompted me to undertake all this work, which seemed to me destined to bring about a revolution in . . . chemistry.'

Antoine Lavoisier, 1773

Lavoisier's contribution to chemistry went beyond setting out a new notion of the elements. In the 18th century, the air around us came in for close scrutiny, and the results did, as Lavoisier predicted, bring about seismic changes.

Solid carbon dioxide rapidly vaporizes at room temperature.

The invisible air

The existence of solids and liquids and the differences between them are obvious, even to the most casual observer. Movement between the two states had been familiar for millennia: people could watch ice melting and water freezing, and many activities, from cooking and metalworking to glass-blowing, made use of the changes of state from solid to liquid and back again. Gases, however, are less immediately obvious. For one thing, most gases are invisible. As even Aristotle noticed, some liquids evaporate, producing the 'exhalation' that he detected from wine. Vapours either condense or dissipate, they can seem to pass into nothing, and substances which are a gas at room temperature are often colourless and odourless.

We can't see air but we can see its effects in blowing wind.

Melting ice is a change of state that has always been familiar to humans.

Jan Baptist van Helmont, the Flemish chemist who grew a willow sapling in a pot, missed a key component of the raw material that was incorporated into his tree as it grew: he took no account of what the tree might have gained from the air. Yet it was he who coined the word 'gas' in the mid-17th century, adapting it from the Greek 'chaos', often used to mean 'void'.

Gases are a tricky concept. When we inhale we obviously breathe in something which is not a solid or a liquid and is invisible. If you stick a tube into a bowl of water and blow through it, you make bubbles. So whatever is in your lungs is significant, has volume and, after a fashion, is capable of being measured. But we can't see the air and we notice it only when it is missing or carries a scent or smoke or moves solid objects (in a gale, for example).

From element to mixture

Air was one of the four or five elements named by ancient cultures. These elements were thought of more as metaphysical principles than indivisible physical substances, but the air was clearly considered one type of matter. The nature of air as a gas or mixture of gases did not come under scrutiny until the 17th century. It was investigated first through its physical properties, treating it as a single substance.

MERCURY SAFER THAN WATER

Galileo and Gasparo Berti had already found that a syphon would not work at a height above 10.3m (34ft) and that water in a closed tube, upended in a bowl of water, would make a column only 10.3m tall. Torricelli was keen to explore the phenomenon, but was already under suspicion of sorcery or witchcraft so he wanted to keep his experiments discreet. For that reason he chose mercury, a much heavier liquid than water, as it would enable him to use a shorter tube. His tube of mercury needed to be only 80cm (31½in) tall, so was easy to keep out of sight.

Working with gases – and without

As most gases are invisible, it was difficult for early scientists to work with, observe and describe them. They had to be investigated by the pressure they exert or the volume they occupy and, later, by their role in chemical reactions.

Torricelli's barometer consisted of a glass tube in a bowl of mercury.

In 1643, an Italian student of Galileo, Evangelista Torricelli, invented a barometer. It consisted of a glass tube sealed at one end and filled with mercury, inverted in a bowl of mercury. It demonstrated for the first time that the atmosphere can exert a pressure, and therefore that air has mass. When the air pressure decreased, the pressure it exerted on the surface of the mercury in the bowl reduced and the level of the mercury in the tube fell, creating a vacuum at the sealed end of the tube. As the air pressure rose, the weight of it pushing down on the surface of the mercury in the bowl forced mercury up into the tube, reducing the amount of space at the top.

Establishing that the invisible air has mass was a very significant first step. The notion that the space above the liquid in the barometer was empty – a vacuum – was just as challenging for many observers. To counter the charge that the empty space in Torricelli's tube might be filled with vapours from the liquid, the French mathematician and physicist Blaise Pascal repeated the experiment in 1646 using tubes of wine and of water alongside each other. If the liquid had been evaporating, the level of the wine should have been lower than that of the water (as it is more volatile), but it was not.

The power of nothing

Boyle was the first person to investigate gases with any rigour. Indeed, his very first experiments were with gases – or, rather, without gases but with the vacuum left by removing them.

With the help of his assistant Robert Hooke, Boyle built a vacuum pump, a device which could extract all the air from a glass dome by means of a piston. (It was not his own invention; he had read in 1657 of a similar 'air pump' produced by the German scientist Otto von Guericke.) Boyle published the results of his experiments with it in 1660 under the title, *New Experiments Physico-Mechanicall, Touching the Spring of the*

WILD HORSES CAN'T PULL THEM APART

Otto von Guericke staged a dramatic demonstration of his air pump and the power of the vacuum it produced. He did this using the 'Magdeburg hemispheres', two copper hemispheres, 50cm (19.5in) in diameter, that fitted together perfectly. When greased, the seal was airtight. At great expense, von Guericke acquired 12 horses and showed that even the power of them pulling together was insufficient to act against the vacuum and separate the hemispheres when the air had been removed from them.

Demonstrating the power of a vacuum – two teams of horses fail to separate von Guericke's Magdeburg hemispheres.

Air, and its Effects. By the pretty phrase 'the Spring of the Air', he meant 'pressure'. Boyle and his colleagues at the newly founded Royal Society experimented with the vacuum pump to discover the effects that removing air would have on the mercury in a thermometer, on a burning candle and (inevitably) on a living mouse.

Boyle was certain that when he pumped the air out of the glass dome, he produced a vacuum – but not everyone was convinced. Since Aristotle had pronounced on the impossibility of a total void nearly 2,000 years previously, there had been considerable reluctance to accept that there could be a space where nothing at all existed. One rationalization put forward by the philosopher Thomas Hobbes (1588–1679) was that when Boyle pulled the plunger of the piston, this increased pressure outside the vessel and tiny eel-like particles of fluid (gas) could slip through the glass wall of the dome. Both Boyle and Hobbes agreed that the difference in the behaviour of solids and fluids can be accounted for by the size and shape of their particles. Clearly the dome did not contain the air that it contained before, but Hobbes felt it must contain *something*. As that something was made of the most refined, tiny, slithery particles, he thought that it must be aether – the super-refined fifth element which Aristotle maintained occupied the celestial spheres and everywhere that was not occupied by something else. Boyle, though, trusted the evidence of his experimental work and continued to maintain that he had created a vacuum – a space that contained nothing at all.

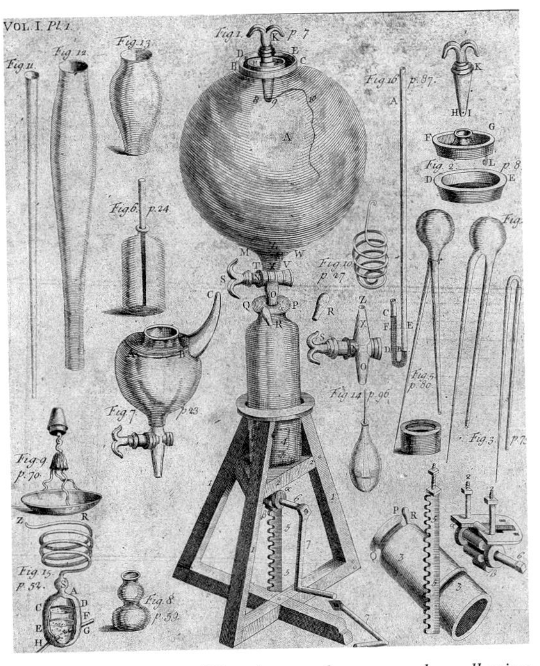

Boyle's first air pump. The sphere at the top was glass, allowing experiments in the vacuum to be observed.

WHERE DOES IT GO?

People objected to the idea of a vacuum for various reasons. Some found it inconceivable that there could be a fixed volume that contained nothing – how can 'nothing' have dimensions? Others felt the outside universe would have to expand by the volume of the emptied dome and this was ridiculous. Hobbes was worried about the entire idea of inexplicable invisible things, and wanted to keep mysterious entities out of science. He maintained that there is no incorporeal substance of any kind, arguing that even God is corporeal.

Interest in gases spread beyond the laboratory. The Montgolfier brothers exploited the behaviour of gases to lift their hot air balloon in 1782.

The 'spring of the air'

Boyle experimented with the pressure and volume of air, formulating Boyle's law in 1662. This finds an inverse relationship between the pressure and volume of a gas at a constant temperature. The pressure exerted by a fixed mass of gas at a constant temperature in a closed system decreases as the volume increases and vice versa. Boyle's law can be expressed as:

$$PV = k$$

where P is pressure, V is volume and k is a constant: the pressure multiplied by the volume will remain the same in a closed system. The law can also be expressed as:

$$P_1V_1 = P_2V_2$$

to compare the same mass of gas in two different sets of conditions. It might seem that this is more physics than chemistry, but it was an essential starting point for pneumatic chemistry.

More than one air

It was far from apparent that there is more than one type of gas in the air – or even that there could be more than one gas at all. It seems strange that although people were aware from early times that noxious 'air' in mines could cause poisoning, they did not come up with a theory of separate gases. Even when it became obvious that more than one type of gas exists, the types were considered variants of air.

The first person to suggest that air has more than one component was the Italian polymath Leonardo da Vinci. He noticed that air is in some way consumed by respiration and also by a burning candle – but it doesn't all disappear. His conclusion was that air contains at least two ingredients. As usual, da Vinci's work was preserved in his notebooks rather than published, so did not contribute to the intellectual debate.

The Polish alchemist Michał Sedziwój, or Sendivogius (1566–1636), also discovered that air is not a single substance and includes a life-giving component. This 'food of life' he identified as the same gas as that given off by heating saltpetre (potassium nitrate), a discovery he revealed in his book *New Chemical Light* in 1604. It was 170 years before mainstream European chemists would make the same discovery, isolating and naming oxygen.

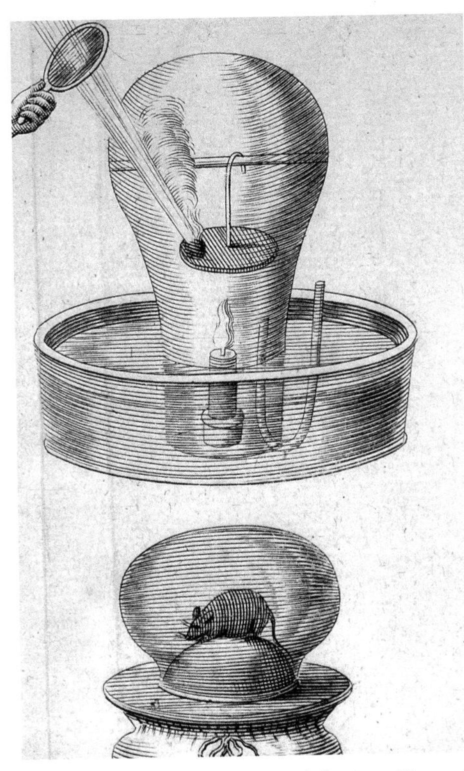

Mayow's experiment demonstrated that breathing and burning use up the same component of the air.

The food of life

Van Helmont, like Leonardo and Sendivogius before him, discovered that air is more complex than it appears. He set a candle burning in the middle of a wide tray of water and upended a glass flask over the candle, its mouth resting in the water. After a while, the candle went out. The level of the water inside the flask rose, which van Helmont explained by suggesting that some part of the air had been consumed and the space it had occupied was now filled with water that had flooded in to replace it.

Van Helmont's experiment was repeated and investigated for more than 150 years. In 1674, the English physician John Mayow (1641–79) used a variant of it to demonstrate that only a part of the air is combustible or used in respiration, but this time the investigation was quantitative. He put a mouse or a burning candle in a closed container over water and waited until the inevitable end came, then measured the rise in the water level. He found that about one-fourteenth of the volume of the air had been removed by the candle or the mouse, demonstrating not only that a candle and a mouse use the same component, but that air is made of more than one type of gas and one type represents around a fourteenth of the whole composition. He named the part that was consumed, rather confusingly for us, *spiritus nitroaereus*.

Sendivogius' 'food of life' was finally isolated by the Swedish apothecary Carl Wilhelm Scheele in 1772 or 1773 and the English chemist Joseph Priestley in 1774. Scheele found that heating manganese oxide until it was red hot produced something he called 'fire air' because of the brilliant

People have exploited fire since prehistoric times, but understanding it proved a challenge.

sparks it made on contact with hot charcoal dust. He discovered that he could make the same 'fire air' by heating potassium nitrate, mercury oxide, or indeed many other substances. Unfortunately, although he made careful notes of his experiments, Scheele delayed publication. Priestley released his results first and is generally credited with isolating oxygen and linking it with combustion and respiration.

Fire and air

Of the ancient elements, air, water and earth are easily associated with genuine elements: the gases, liquids and solids (predominantly minerals and metals) found around us. But fire turned out to be a puzzle for the early chemists. It is a transformative agent, it can't be isolated on its own and is always found acting on other matter. Chemists began thinking more scientifically about fire in the 16th century.

Putting on weight

It is clear that things that are burned are usually diminished. They might be reduced to ashes and generally weigh less than they did before burning. This suggests that something is lost during burning. There was the curious case of metal calces (oxides), though. Some metals form oxides when burned, so end up weighing more than before. The Italian physician Giulio Cesare della Scala suggested in 1557 that the weight gain of burning lead and rusting iron might be related, and that both are caused by the absorption of some particles from the air. But there was no theoretical framework to explain this, so it went no further.

Boyle suggested instead that something from the flame was added to the metal when it burned: '[the] flame itself may be, as it were, incorporated with close and solid bodies, so as to increase their bulk and weight.' He weighed metal, sealed it in a glass container and exposed it to flame. Afterwards, he broke the container and weighed the contents, finding that the mass had increased. He suggested that 'igneous particles' from the flame could penetrate the glass vessel.

Fiery phlogiston

Georg Stahl (1660–1734) developed this idea into the theory of phlogiston. Phlogiston, he claimed, is a very subtle (fine) principle that combines with matter and is present

Stahl's theory of phlogiston led chemists up the wrong path.

> *'That phlogiston should communicate absolute levity to the bodies with which it is combined, is a supposition that I am not willing to have recourse to, though it would afford an easy solution of the difficulty.'*
>
> Joseph Priestley, 1774

in all flammable materials; it is released on burning. Its name comes from the Greek, meaning 'burning up'. The more phlogiston a substance contains, the less residue remains when the substance is burned. So something like paper, which reduces to fine ash, contains a lot of phlogiston.

Stahl believed that phlogiston:

- gives matter the quality of flammability
- is released into the air when matter is burned
- cannot be detected on its own
- gives fire the power of motion, and that motion is circular
- is the foundation of colour
- cannot be destroyed and it can't escape from the atmosphere, so the fixed amount of phlogiston is constantly recycled
- is needed for something to burn, and so also is air
- is abundant in highly flammable substances such as oil.

He made another point, which seems odd to us: heating a metal calx *adds* phlogiston, so restores the original metal. He does not seem to have thought it had mass and speaks of it only as a 'principle' and never as a form of matter.

German physician Johann Juncker, noting that when a metal is burned its mass increases (as a calx is formed), suggested that phlogiston could have negative mass. Priestley, among others, was not convinced by this last argument, though he did accept the general principle of phlogiston.

Phlogiston was soon implicated in respiration as it was clear that burning objects and respiring mice had a similar relationship with air. The theory was that as flammable things contain phlogiston, burning them releases phlogiston into the air; respiring plants and animals seemed to do the same thing. It was not thought – as we now know is the case – that something is used up in the air but that something was released into it. The model

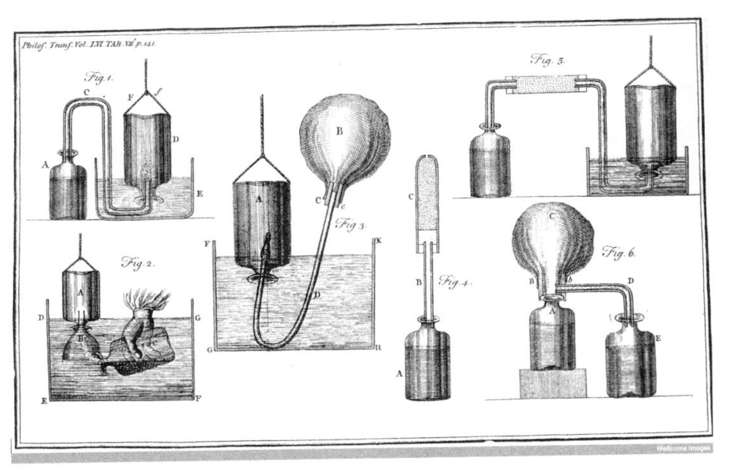

Equipment devised by Cavendish (see page 97) for trapping, moving and measuring gases produced in experiments described in papers he published in 1766.

was entirely the wrong way round, and it took some unravelling. This was despite Mayow and van Helmont suggesting the correct situation, that breathing and burning take the same thing from the air. Neither view gives a complete account, of course: something is used up and something is added to the air – but it is not phlogiston.

PRODUCTIVE MISTAKES

Priestley couldn't find anyone to illustrate one of his works, so he taught himself perspective drawing. In the process, he found that India rubber can be used to erase pencil marks. He recorded the discovery in the preface to the book.

JOSEPH PRIESTLEY (1733–1804)

Joseph Priestley has the unusual distinction of being the only chemist to have riots named after him. They were the result of his unpopular politico-religious beliefs and led to him fleeing England for America in the 1790s.

Born into a middle-class household in Yorkshire, England, Priestley was formidably clever as a young child. His family hoped he would become a Calvinist minister and he learned Latin, Greek and Hebrew, later adding French, Italian, German, Arabic and Aramaic. A serious illness ended both hopes of a ministry and Calvinism: he was left with a stutter and became a Unitarian after despairing of having the type of spiritual experience that would confirm he was one of the elect. His strong religious feelings led him to write theological tracts; his interest in chemistry ran alongside his philosophical pursuits for his whole life.

His first scientific work was with electricity, encouraged by Benjamin Franklin, but he turned to working on gases in the 1770s. Priestley built a 'pneumatic trough', a device first described by the botanist Stephen Hales in 1727, which allowed him to collect gases over water as they were produced by reactions. He identified eight new gases, more than anyone else has ever achieved. He was a supporter of the theory of phlogiston, which led to arguments with Lavoisier.

Priestley supported both the French and American revolutions, anticipating that they would lead to the overthrow of all earthly regimes and hasten the coming of the Millennium of Christ foretold in the Bible. This view was unpopular, and in 1791 his house and laboratory were destroyed by an angry mob in four days of riots often referred to as the Priestley Riots. In America, he tried to set up a model community in Pennsylvania. Although his utopian dream was not realized, he did build a luxurious home with a laboratory.

Ernest Board's painting of 1912 imagines the moment Priestley, playing backgammon, hears that rioters are approaching his house.

'Inflammable air'

There was even a candidate for phlogiston.

The first gas to be isolated was probably one that does not occur naturally in the atmosphere on its own in any noticeable quantity – hydrogen. Paracelsus probably found it in the 16th century, for he produced something he called 'inflammable air' by experimentation. Boyle was the first person to document the generation of hydrogen in a repeatable experiment, though, in 1671. He discovered that iron filings in a dilute solution of acid produce bubbles of a gas that he could trap in an upturned tube of water as they emerged. Even so, it was left to the English chemist Henry Cavendish (1731–1810), nearly a century later, to recognize that it is a discrete substance, different from air. In 1766, he called it 'inflammable air', and suspected it might be the same as phlogiston. Lavoisier named it 'hydrogen' in 1783.

Burning and breathing

That breathing and burning had similar effects on air suggested a link between them, and chemists were quick to explore it. Alongside Lavoisier, the person most involved in investigating gases in the 18th century was Joseph Priestley.

Fracturing the air

In 1774, Priestley discovered that if he placed a lump of mercuric oxide in a sealed container and then heated it with sunlight magnified through a lens, it produced a gas that was 'five or six times as good as common air' at sustaining a candle or a live mouse. That is, either would continue to burn/breathe for five or six times as long as they would in normal air.

He identified this super-duper air as 'dephlogisticated air' because he believed it to be air from which all the phlogiston

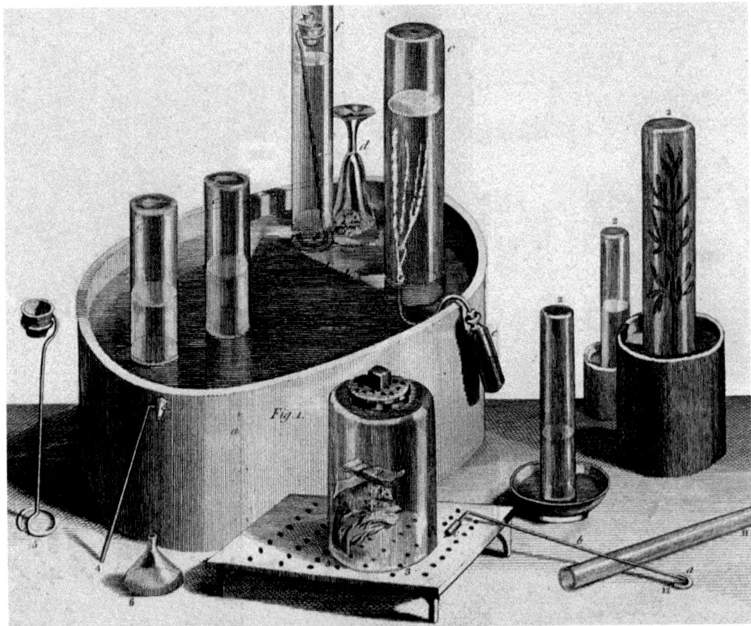

Pneumatic trough and other equipment used by Priestley.

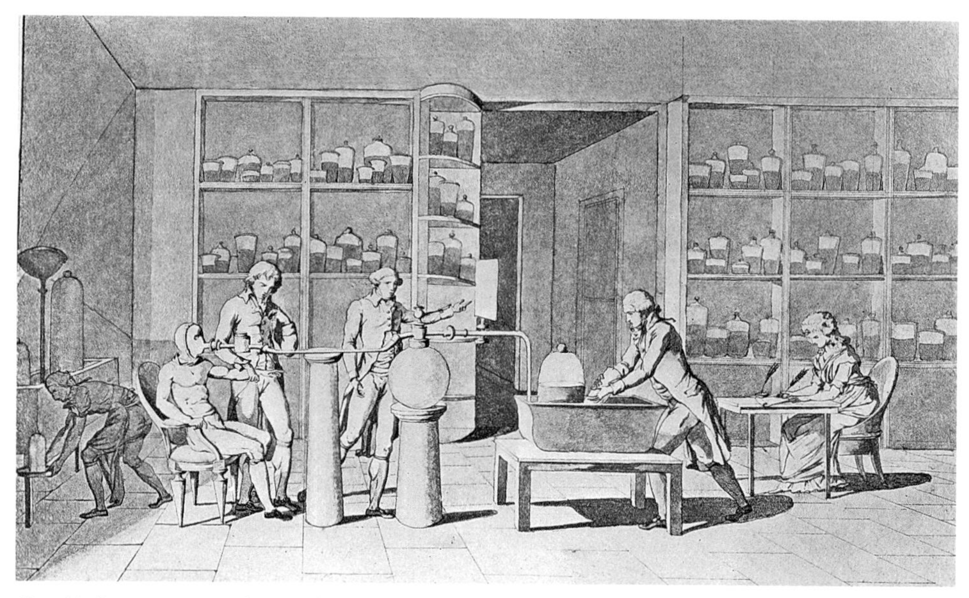

Lavoisier's experiments with gases of respiration were carefully recorded by his wife, who has included herself in this picture (far right).

had been removed. As it contained no phlogiston, it could (he contended) sustain a lot of burning or breathing before becoming saturated. Once air was saturated with phlogiston, it could hold no more so respiration or combustion could not continue – the mouse would die, the candle would go out.

The implications were immense: 'Air is not an elementary substance, but a composition,' Priestley wrote, overturning the notion of air as an element that had gone virtually unchallenged for more than 2,000 years. Air contained, at least, its breathable portion and some phlogiston.

Priestley, inevitably, inhaled some 'dephlogisticated air' and reported good effects: 'The feeling of it in my lungs was not sensibly different from that of common air, but I fancied that my breast felt peculiarly light and easy for some time afterwards. Who can tell but that in time, this pure air may become a fashionable article in luxury. Hitherto only two mice and myself have had the privilege of breathing it.'

Priestley visited Europe and met Lavoisier, who he told about his experiment with dephlogisticated air. Lavoisier had been the student of Guillaume-François Rouelle, who had introduced Stahl's phlogiston theory to France, so was already familiar with it – but he was not a fan. Neither burning nor respiration release phlogiston, he argued, but both require the special 'air' that Priestley had isolated. Lavoisier proposed that air has two normal ingredients (neither of them as unorthodox as phlogiston). He suggested that one of these is essential to breathing and combines with metals; the other causes asphyxiation

Priestley's musing that 'in time, this pure air may become a fashionable article in luxury' finds realization in the trendy 'oxygen bars' of the 21st century.

and can't support burning. The first he described as 'eminently respirable' and explained that it combines with a metal or organic substance in combustion. Two years later, in 1776, he named it *oxygène*, from Greek words meaning 'acid generator' (because he believed – wrongly – that all acids contain oxygen).

In 1783, Lavoisier roundly denounced the phlogiston theory, saying phlogiston was 'imaginary' and 'a veritable Proteus that changes its form every instant'. He was aware that his rejection of it would be unpopular and acknowledged that his contemporaries 'adopt new ideas only with difficulty'. Still, he was happy to report in 1791 of his alternative explanation that, 'All young chemists adopt the theory.'

After finding that either a breathing mouse or a burning candle could exhaust

BREWERS AND BUBBLES

In 1767, Joseph Priestley was appointed minister of a parish near Leeds in England. He spent a lot of his time at the local brewery investigating the brewing process; it produced a gas that he was keen to collect and use in his experiments – the 'fixed air' described by Joseph Black (carbon dioxide). Eventually, he was banned from the brewery after inadvertently polluting a vat of beer with some of his chemicals – but not before he had discovered how to make soda water. Priestley published his instructions in 1772 in a pamphlet entitled *Directions for Impregnating Water with Fixed Air.* He made no attempt to commercialize his invention, but Johann Jacob Schweppe did, eventually making a fortune from it through the sale of carbonated (fizzy) water.

Bubbles bonanza: a Schweppes advertisement from the 1930s.

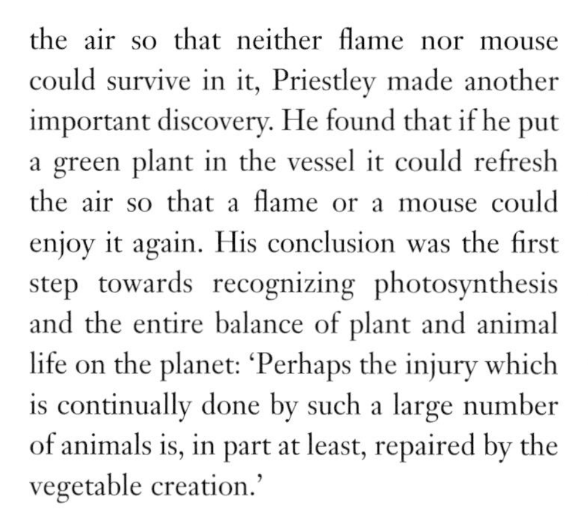

the air so that neither flame nor mouse could survive in it, Priestley made another important discovery. He found that if he put a green plant in the vessel it could refresh the air so that a flame or a mouse could enjoy it again. His conclusion was the first step towards recognizing photosynthesis and the entire balance of plant and animal life on the planet: 'Perhaps the injury which is continually done by such a large number of animals is, in part at least, repaired by the vegetable creation.'

Carbon dioxide unravelled

Perhaps surprisingly, carbon dioxide was discovered before oxygen. Although carbon dioxide makes up only a small portion of the atmosphere, it is readily produced in quite simple experiments and procedures, which is how the early chemists came by their gases.

Carbon dioxide was discovered by the Scottish chemist Joseph Black in 1754. He found that if he heated calcium carbonate a heavy gas was produced that could not sustain a flame or a breathing animal. He called it 'fixed air' because it can be 'fixed' (absorbed) by strong bases (the chemical opposite of acids).

Carbon dioxide has more uses in the world than making fizzy drinks. The most important of these was uncovered by the Dutch physiologist and chemist Jan Ingenhousz, who repeated Priestley's mouse experiment in 1778. Ingenhousz demonstrated that a plant needs sunlight in order to work its magic on the air. Nearly twenty years later, in 1796, the Swiss botanist Jean Senebier showed that in the presence of sunlight a green plant takes in carbon dioxide and releases oxygen.

And the rest . . .

The greater part of normal air, making up about 78 per cent of Earth's atmosphere, is nitrogen. It was discovered in 1772 by the Scottish physician Daniel Rutherford, who was Black's doctoral student. Rutherford used three methods to 'phlogisticate' air: he enclosed a mouse in a jar until it died, he burned a candle in the jar until it went out, or he burned phosphorus in the jar until it would no longer burn. In each case, he passed the resulting gas through lime-water to remove the 'fixed air'. The remaining gas, he found, would not support a candle or another hapless mouse. Unlike carbon dioxide, though, it was not soluble in water or in alkalis (solutions of bases). He named it 'noxious air' and reported that it was lighter than normal air and could not be further broken down by 'any other cause of the diminution of air that I am acquainted with'.

It seems that Scheele also discovered nitrogen in 1772, but did not publish his findings until 1777 (he had made the same mistake with his discovery of oxygen, but didn't learn from the unhappy experience). He called it 'spent air' and found it to occupy about two-thirds to three-quarters of the volume of the original air he started with.

The world appears to have been full of people experimenting with gases and not publishing their results as Cavendish, too, apparently discovered nitrogen – or 'burnt air' – some time before 1772. He passed air repeatedly over red-hot charcoal to remove the oxygen, then bubbled it through a

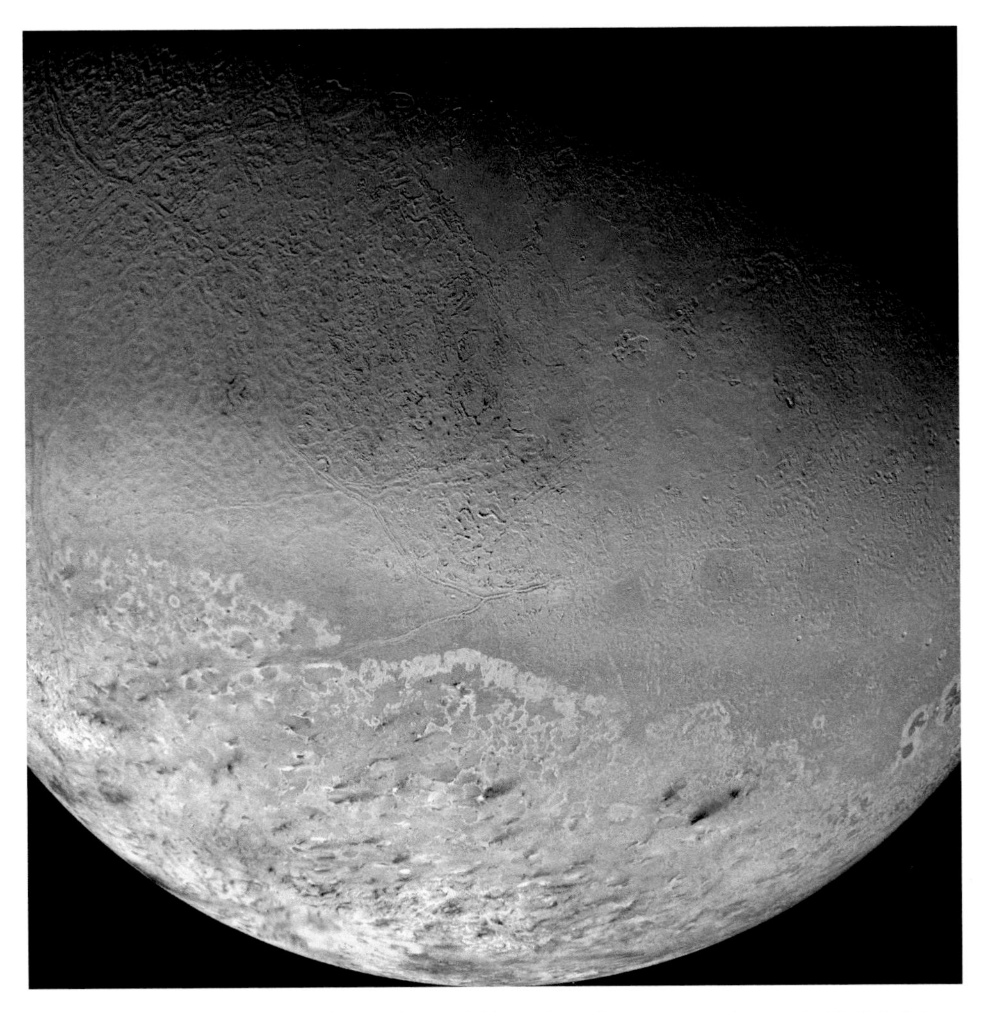

Nitrogen is a gas on Earth, but on Neptune's moon Triton, where the temperature is around -236 ° C, it is a solid. The crust of Triton is 55 per cent nitrogen ice.

solution of potassium hydroxide, removing the carbon dioxide. He found his 'burnt air' to be very slightly lighter than 'common air' and unable to support a burning flame.

Compound or mixed airs?

The next question to ask was whether the different airs in the atmosphere were simply mixed together or formed one or more chemical compounds. It was not immediately apparent how chemists would be able to tell the difference.

Repulsive gases

The English chemist John Dalton (1766–1844) developed an early interest

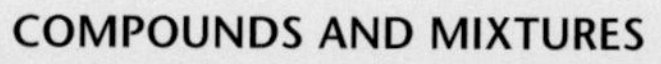

COMPOUNDS AND MIXTURES

Chemists recognize two different ways of putting materials together.

In a compound, chemical bonds are formed between atoms, making a new substance from one or more original substances. The new substance is homogenous – it is made of molecules that are all the same. For example, when sodium and chlorine combine, the result is new molecules of sodium chloride, or salt, which has chemical properties of its own, distinct from those of the separate components.

In a mixture, two (or more) substances are mingled but do not react with one another. No new chemical bonds or molecules are made. The substances can be separated (though sometimes with practical difficulty). An example would be mixing iron filings and sand together. No new compound is formed, and the iron filings could be removed again, unchanged, using a magnet.

in meteorology and kept records of the weather for five years. This led him to think about gases. He argued that air is a mix of gases, that each exists in its own free state and the gases of the air are not chemically combined.

He wondered why, if the gases are only mixed together, they don't separate out into layers with the heavier gases lying closer to the ground and the lighter ones floating above them. His solution was ingenious. He concluded that the particles of a gas repel one another, so will spread out as far as possible, supporting Newton's and Boyle's findings relating to pressure and volume. But instead of having all particles repel all others, he proposed that each type of gas particle repels only its own type. The result would then be a nice homogenous mix of gases, regardless of the mass of the particles – which is exactly what we see.

This conclusion led Dalton to his theory of partial pressures, which states that the total pressure exerted by a mix of gases is the sum of the pressure exerted by all the different gases in a mixture. As different gases have different partial pressures, which suggests some repel their particles more vigorously than others, Dalton had to conclude that the particles are different sizes. This led to his insight that the atoms of each element are unique. And this, in turn, formed the bedrock of his atomic theory (see page 116).

Not all air

Although the gases that first attracted attention were, predictably, those of the air, it emerged in the 19th century that they were certainly not the only gases. There are other gaseous elements, and many compound gases.

Halogen gases

The Swedish chemist Carl Scheele discovered chlorine in 1774 when he combined hydrochloric acid and pyrolusite (manganese dioxide, MnO_2). He thought the gas was a compound containing oxygen, but Humphry Davy began investigating it in 1807 and found it to be an element (see page 182). Davy also proposed the existence of fluorine, as did French physicist André-Marie Ampère, but was unable to extract it. As the most reactive of all elements, fluorine is very difficult to isolate. Many chemists were injured trying to extract it from hydrofluoric acid (including Davy himself, Gay-Lussac, Thénard, and Thomas and George Knox). At least two were killed (Paulin Louyet and Jérôme Nicklès). It was finally isolated after much effort (and some injuries) by Henri Moissan in 1886 – work for which he received 10,000 francs and a Nobel Prize (in 1906).

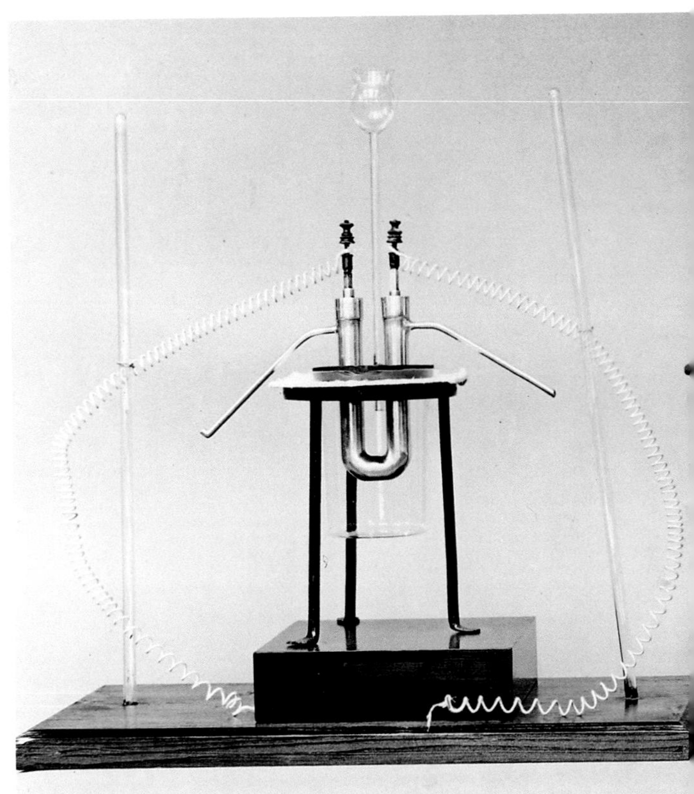

A replica of Moissan's equipment for extracting fluorine by the electrolysis of a solution of potassium hydrogen, fluoride.

Rare gases

The rare or noble gases form the final column of the Periodic Table (see page 130). Called 'noble' because they are largely inert, so seldom react with other elements, they were discovered by Scots chemist William Ramsay at the end of the 19th century.

When Cavendish isolated nitrogen, a tiny bubble remained in his collection apparatus that was not nitrogen. It was not very big, and its presence was ignored for around a century. Cavendish also found that the nitrogen left in the air when everything else was removed was very slightly denser, by about 0.5 per cent, than the nitrogen derived from chemical reactions. Lord Rayleigh, professor of experimental physics at the University of Cambridge, made the same discovery in 1894 and engaged the help of Ramsay to investigate. They passed atmospheric nitrogen over red-hot magnesium, causing the nitrogen to react with the magnesium forming magnesium nitride and leaving the extra component. They found they had isolated a gas that was so unreactive it would not even respond to fluorine; Ramsay described it as 'an astonishingly indifferent body'. They announced it as a new element in 1895, calling it argon from the Greek for 'idle'.

GASES CAN BE FUN

Joseph Priestley had discovered nitrous oxide (N_2O) in 1772, making it by allowing nitric oxide (NO) to lie in contact with iron filings and water:

$$2NO + H_2O + Fe \rightarrow N_2O + Fe(OH)_2$$

In 1798, the young English chemist Humphry Davy was employed to investigate nitrous oxide and its possible uses. He published his long text on the gas and its history in 1800 when he was only 21 years old. In it, he remarked, 'As nitrous oxide appears capable of destroying physical pain, it may probably be used with advantage during surgical operations in which no great effusion of blood takes place.' Sadly, the recommendation was ignored for more than 40 years, until 1844. Pain relief was not Davy's concern; he was interested in chemical bonds. But that didn't stop him enjoying the fruits of his research. He and his friends used nitrous oxide recreationally, inhaling it from oiled silk bags at parties. It was also used in stage shows. *The Times* of 1819 reported of one such show: 'The nitrous oxyd, or laughing gas was inhaled by a gentleman who after laughing sprung up in the air to the astonishing height of six feet from the ground.'

Nitrous oxide made parties fun in the early 19th century.

Ramsay pointing out the column of rare gases in the Periodic Table.

Ramsay went on to find helium, which had previously been identified in the Sun by spectroscopy (see page 183) in 1868 but never before found on Earth. Ramsay then felt sure there was a whole column of rare gases to be discovered, but met with hostility from other chemists. Rayleigh withdrew under the onslaught but Ramsay persisted and was rewarded not only with more noble gases (neon, xenon, radon and krypton), but with a Nobel prize in 1904.

The surprise of water

It was already known that gases could form solid compounds with solids – the formation of calces was evidence of that. But in 1781

> **RICH AND RECLUSIVE**
>
> Cavendish was an odd character. Born wealthy, and dying even wealthier, he indulged in none of the pleasures or peccadilloes that would have been available to an 18th-century gentleman. Instead he turned over his entire house to the pursuit of chemistry, making his drawing room his main laboratory and other rooms lesser labs, workshops, a forge and an astronomical observatory. He shunned company, and if anyone came uninvited to visit him they were served simply a leg of mutton and nothing else – whether in the hopes of driving them away or because he had not the imagination to think of anything else is unclear. He refused all requests to paint his portrait and only one rapidly and furtively acquired sketch survives. He died alone, having told his servant, 'I am going to die. When I am dead, but not till then, go to Lord George Cavendish and tell him—Go!' His enthusiasm for chemistry, and his meticulous quantitative methods, led him to the discovery of hydrogen, the composition of water and the first hint at the presence of the noble gases in atmospheric air.

Henry Cavendish discovered that if he set light to 'inflammable air' (hydrogen) in the presence of ordinary air, a small amount of 'dew' formed on the walls of the glass vessel. This, he concluded, seemed to be ordinary water. It's hard to imagine now how revolutionary and surprising this finding must have been. Not only was water not, after all, an element – it was also a liquid made by combining two gases.

Cavendish explained this by saying that one gas is overly-phlogisticated water and the other dephlogisticated water, and that combining them produces properly phlogisticated water. Lavoisier's simpler (and correct) explanation was that water is composed of the two gases. As combustion consists of adding oxygen, oxygen is one half of water and 'inflammable air' is the other half. The entire reaction was explicable without resorting to phlogiston.

Cavendish reported that two parts of hydrogen combine with one part of oxygen. The proportions were demonstrated conclusively by the Polish chemist Johann Ritter in 1800 after using electrolysis to break water into its constituent gases.

Back to gases

The study of gases had begun with investigations of the gaseous state rather than individual gases. It returned to that at the start of the 19th century, with work that would tie gases in with atomic theory, the subject of the next chapter.

Up and up

The French chemist Joseph Gay-Lussac (1778–1850) took over from where Boyle had left off in his work on pressure. In 1801–2, Gay-Lussac made extensive studies of the behaviour of gases and concluded that they all increase equally in volume with equal rises in temperature. This is a slightly surprising finding (or it was if you didn't know just how much of a volume of gas is empty space), but it had actually been discovered 15 years previously by Jacques Charles, who had not published it. It's now known as Charles' law.

CHAPTER 6

ATOMS, ELEMENTS AND AFFINITIES

'It is certain that all bodies whatsoever, though they have no sense, yet they have perception; for when one body is applied to another, there is a kind of election to embrace that which is agreeable, and to exclude or expel that which is ingrate; and whether the body be alterant or altered, evermore perception precedeth operation; for else all bodies would be like one to another.'

Francis Bacon, 1620

By the end of the 18th century, Lavoisier had defined elements as those substances that cannot be further broken down, and Boyle had spoken of simple 'bodies' that are the equivalent of modern atoms. But the two had not quite come together.

Joseph Wright of Derby's Experiment on a Bird in the Air Pump *(1768) shows Robert Boyle removing the air from a glass dome containing a bird. Boyle's objective was to show that a vacuum could be cold.*

Atoms and elements

In modern chemistry, the notions of elements and atoms are inseparable. Like the idea of elements, that of atoms was first proposed by the Ancient Greeks around 2,500 years ago.

Ancient atoms

The atoms proposed by Leucippus and Democritus (see page 21) were like 'modern' atoms in some ways and unlike them in others. They were said to be infinite in number, and to come in various sizes and shapes. They were solid, with no internal gaps, and could not be further divided (their name means 'uncuttable'). They move around in the infinite void. If they come into contact with others, they either repel them or collide and become entangled, forming clusters held by hooks or barbs on their surface. They can't be destroyed or generated, but are unchanging and eternal. All the changes we see in the world around us are produced by atoms changing places and interacting with one another in different ways.

The idea of atoms hooked together is the first model of an atomic bond. It appealed to the Roman philosopher Lucretius (99–*c.*55BC) who described hard and dense materials as being made of atoms that are 'hooked together, and must be held in union, because welded together through and through out of atoms that are,

Lucretius suggested the atoms of substances such as olive oil might be large or easily tangled, accounting for the liquid's viscosity.

as it were, many-branched'. He felt that most liquids are formed of smooth, round particles that easily flow and slide over each other, but sluggish fluids such as oil might have atoms that are 'larger or more hooked and intertangled'.

Atomic shapes

Plato associated each of the four elements with a perfect geometrical 3D shape (the Platonic solids), which helped to explain the characteristics and behaviour of the element. Fire was identified with the tetrahedron, a spiky shape that would produce the painful sensation of burning; air was octahedral as its many small faces make it closest to a sphere and enable particles to roll and glide easily; water he thought icosahedral, the next most nearly spherical and therefore able to flow; and earth he thought was made of cubes, which can pack densely but also separate easily, as do particles of crumbling soil.

Plato's model also explained the possible transformation of the elements. As the faces of the tetrahedron, icosahedron and octahedron are all equilateral triangles, it's possible to mash, break and recombine the elements fire, air and water. For example, fire particles can combine to make air particles, or an air particle can break into fire particles. Only the cube does not have triangular faces and so earth can't partake in any transformations.

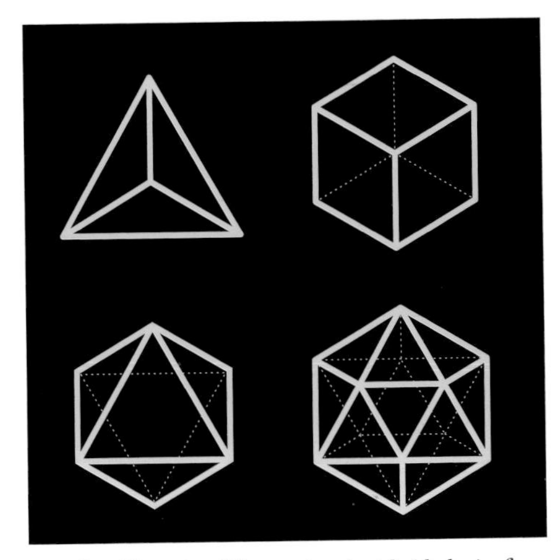

The four Platonic solids, associated with (clockwise from top left) fire, earth, water and air.

Atoms in France

No one seems to have thought much about atoms between the time of the Ancient Greeks and the 17th century. Then two French philosophers, René Descartes (1596–1650) and Pierre Gassendi (1592–1655), reopened the can of worms – or eel-like particles, as Descartes thought of them.

As we have seen, Aristotle had supposed some form of primary matter is given 'substantial form' through a series of properties. This idea was developed in the European universities of the Middle Ages and was still the prevalent view when Descartes and Gassendi came, separately, to think about the nature of matter. It held that matter has no particular qualities of its own, and attributes such as shape, colour and texture are bestowed by the form it takes. So matter incorporated in a feather will take on different attributes from matter incorporated in a muddy puddle or a table. Descartes rejected this view: 'If you find it strange that I make no use of the qualities one calls heat, cold, moistness, and dryness . . . as the philosophers do, I tell you that these qualities appear to me to be in need of explanation.'

Instead, he felt it was possible to explain all inanimate objects by qualities that could

be empirically determined: 'motion, size, shape, and the arrangement of their parts'. Even so, he was not stepping very far from the methods of the scholastics. Like them, he began by constructing a metaphysical model and then looked to the physical world for evidence to support it – the opposite of the scientific method. Descartes even criticized Galileo for proceeding 'without having considered the first causes of nature, [he] has merely looked for the explanations of a few particular effects, and he has thereby built without foundations'.

René Descartes' contributions to intellectual discourse ranged from maths and astronomy to philosophy.

Descartes thought matter could be infinitely divided, and that it forms a continuum, leaving no space. And yet his model did have particles. The Ancient Greeks, who believed in the concept of particles, had to accept a vacuum in which they could move. Descartes, however, wanted to have his cake and eat it: he believed in the notion of a universe jam-packed full of matter but loose enough to allow tiny particles to move through it. Think of a pond with fish and tadpoles. The matter is continuous and some bits – the fish and tadpoles – can move through the rest of it (water), which closes immediately behind them, thus ensuring there are no gaps.

Gassendi was a more traditional atomist, and also came up with the concept of molecules. He defended the idea of the void using the arguments of the ancients, but also by reference to the empirical evidence of his day, including the barometer, developed in 1643 (see page 89). Much of his argument for the existence of atoms is based in the tradition of philosophical debate, but his conclusions still stand today: that all matter must share some essential fundamental features and that atoms provide those features.

In a line of argument developed from Lucretius, Gassendi suggested that atoms must be hard (solid). If they were soft, there could be no hard objects. On the other hand, if atoms are hard, matter with less densely packed atoms can be soft as there is space between the atoms to allow the object to give way. This variable architecture of

atoms also allows substances to be permeable. Ultimately, he was less concerned with the argument of whether or not atoms actually exist than he was persuaded that the assumption of atoms existing was the best possible working hypothesis.

To provide the variety of matter we see around us, Gassendi postulated a variety of atoms. In his view, atoms come in a limited number of sizes and weights, but a great variety of shapes. Atoms have a natural tendency to move, a result of their having weight. Somewhere along the line, matter goes from a collection of agile and active atoms to the stationary objects around us. In up-sizing, they have somehow developed inertia, as objects are not prone to moving without good cause. Gassendi is quite eloquent on the movement of atoms; they are able to 'disentangle themselves, to free themselves, to leap away, to knock against other atoms, to turn them away, to move away from them, and similarly [have] the capacity to take hold of each other, to attach themselves to each other, to join together, to bind each other fast'. In this last he was proposing something like molecules.

Gassendi made Aristotle's properties of hot, cold, wet and dry into four different types of atom. The particles of heat were small and round and the cold particles were spiky pyramids, which he thought explained why cold could feel sharp. He saw light/heat, sound and magnetism as all atomic. He thought the atoms of light or heat (which he considered equivalent) moved faster than other atoms. His explanation of evaporation was that the space between atoms in the liquid increases – an explanation that a modern chemist would accept. But Gassendi thought this was achieved by the heat atoms removing some of the substance

The alchemist Robert Fludd, writing of the creation of the Earth, suggested that heating crystal produced 'a million sensible Atoms flying in the air'. He considered the Creation explicable in terms of chemistry. Although he believed still in the original Greek elements, he thought it 'probable that all things were made of Atoms as some Philosophers have guessed'.

HOT AND COLD ATOMS

In talking of atoms of heat, Gassendi took his cue from Epicurus (341–270BC). He explained heat as the presence of calorific (hot) corpuscles and cold as the presence of frigorific (cold) particles.

The process of freezing was the testing ground for theories of heat in the 17th century. Descartes, with his continuous-matter model, considered that ice forms when the aetherial matter leaves water through its 'pores', seizing it up as a solid. Gassendi explained it as the result of frigorific particles entering the water. The English philosopher Thomas Hobbes thought the particles entering the water were just air, but their presence between the water particles prevented the water moving. Others believed particles of some kind of salt insinuated themselves between water particles, 'fixing them together, like nails'.

To disprove Hobbes' contention that air produces coldness, Boyle carried out a famous demonstration with a bird in a vacuum. Although it's usually thought the point was to show that air could be drawn out of the flask and leave a vacuum – itself a contested thing – it was in fact intended to demonstrate that cold could exist in a vacuum, so there was no need for 'frigorific corpuscles'. (The bird would in any case have fallen to the bottom of the container as it could not fly with no air to give it lift, and it would have suffocated – it was triply doomed.) The callous demonstration finally resolved an argument between Hobbes and Boyle that had been going on for ten years.

atoms, so increasing the ratio of void to matter. He explained that solids dissolve in liquids when the shape of the atoms is such that an atom of liquid and one of solid can lock together.

From atoms to molecules

If there are relatively few types of atoms, they can't account for the variety of matter unless they combine. And if 'principles' are not to account for the properties of matter, something else must. Robert Boyle was a proponent of 'corpuscularism', believing that the nature of – and changes in – matter are the result of particles and their motions rather than the combination of principles. In 1661, he proposed something that approaches the idea of atoms and molecules combining to form compounds: 'Certain primitive and simple, or perfectly unmingled bodies; which not being made of any other bodies, or of one another, are the ingredients of which all those called perfectly mixt bodies are immediately compounded, and into which they are ultimately resolved.'

Getting together

In 1704, Isaac Newton suggested a principle that became very important in chemistry: that 'particles attract one another by some force, which in immediate contact is exceedingly strong, at small distances performs the chemical operations, and reaches not far from the particles with any sensible effect'. This was the first step towards understanding how atoms might combine into molecules, and so how elements might combine to make compounds.

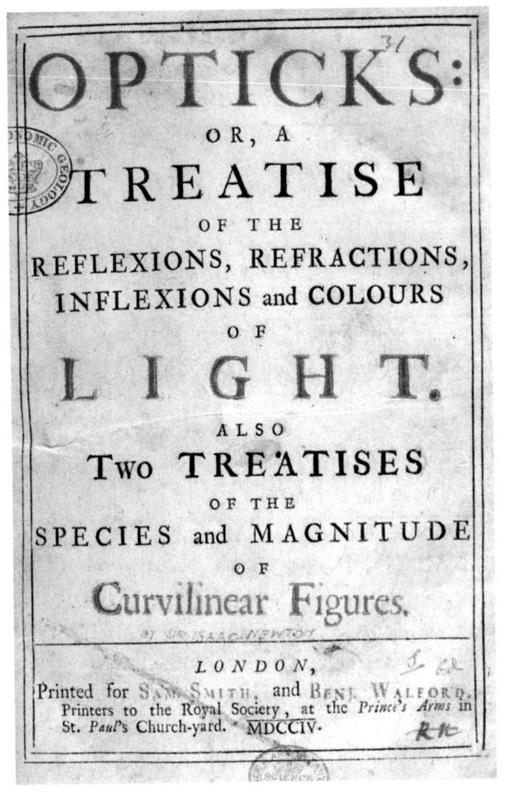

OPTICKS:
OR, A
TREATISE
OF THE
REFLEXIONS, REFRACTIONS,
INFLEXIONS and COLOURS
OF
LIGHT.
ALSO
Two TREATISES
OF THE
SPECIES and MAGNITUDE
OF
Curvilinear Figures.

LONDON,
Printed for SAM. SMITH, and BENJ. WALFORD,
Printers to the Royal Society, at the *Prince's Arms* in
St. *Paul's* Church-yard. MDCCIV.

The title page of Newton's Opticks *(1704) in which he set out ideas about a hierarchy of substances.*

Atoms and their affinities

The notion that there is some form of attraction between atoms of different types would explain why elements come together to form compounds. But Newton had gone further than that. He'd suggested there was a hierarchy within substances, and some would be more willing to form alliances than others. This was picked up by the French translator of *Opticks*, Étienne Geoffroy, who published a table of chemical 'affinities' in 1718 showing the various substances in order of their enthusiasm for reacting together.

> *'And is it not for want of an attractive virtue between the Parts of Water and Oil, of Quick-silver and Antimony, of Lead and Iron, that these Substances do not mix; and by a weak Attraction, that Quick-silver and Copper mix difficultly; and from a strong one, that Quick-silver and Tin, Antimony and Iron, Water and Salts, mix readily?'*
>
> Isaac Newton, *Opticks*, 1704

The first row of Geoffroy's table (see page 114) shows a header species; below it are all the substances that will form an affinity with it (react). A substance will replace any substance beneath it in the column if given the chance. For example, column 9 shows that iron, copper, lead, silver, antimony and mercury will all combine with sulphur, with iron having the greatest affinity. Other chemists refined and expanded the affinity table throughout the 18th century.

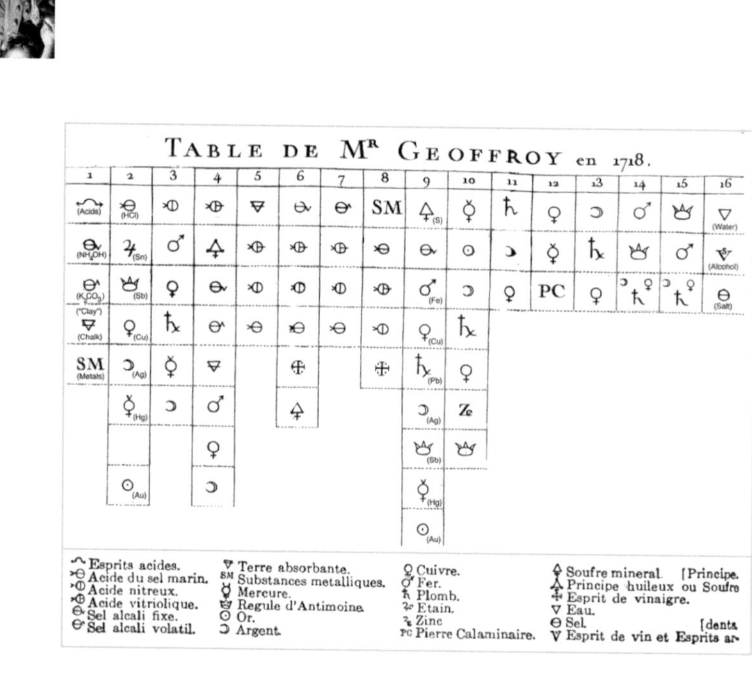

Geoffroy's table of affinities shows the sequence of reactivity of various substances.

From affinity to reaction

The modern notion of a chemical reaction, beginning with a set of reactants and ending with products, came out of the study of affinities and the development of ever more comprehensive affinity tables. The Scots chemist William Cullen first used a diagram to show the progress of a chemical reaction. In a lecture on affinities at the University of Glasgow in 1756, he used brackets and an arrow to show how the hierarchy of affinities led to one chemical replacing another in a compound. In Cullen's depiction of a reaction, the bracket shows the bonded chemicals and the dart (arrow) shows the direction of affinity (see table, opposite top). Cullen's diagrams looked cryptic as he used the alchemical symbols for the substances he was discussing.

As time passed and affinity tables grew larger and more complex, it became apparent that the system was not infinitely extendable. The largest was produced in 1775 by the Swedish chemist Torbern Bergman and was included as a fold-out map in his textbook *A Dissertation on Elective Attractions*. With 59 columns and 50 rows, it covered thousands of possible reactions and was difficult to use. Bergman's table listed 25 acids, 15 earths, and 16 metallic calces; Geoffroy's original table had included only four acids, two alkalis and nine metals.

ANCIENT AFFINITIES

Albertus Magnus had used the concept of 'affinity' in the 13th century to explain the likelihood of substances reacting. He considered that the greater the affinity between substances in terms of their resemblance, similarity of properties or relationship to one another, the greater their tendency to react. The belief that like associates with like goes back to the teachings of Hippocrates, but was first applied to practical chemistry by Magnus.

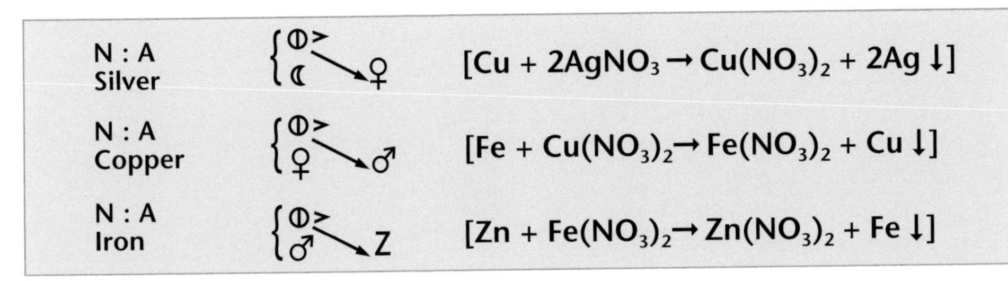

In Cullen's diagram, N:A represents nitrous acid. The sequence shows that in a solution of silver in nitrous acid, copper will replace silver, causing silver to precipitate out (line 1); iron will then replace copper (line 2); and zinc will replace iron (line 3).

Bergman's large and unwieldy table of affinities.

HUMAN AFFINITIES

It doesn't take a great leap of imagination to see the metaphorical potential of the affinity table. The principle that if a chemical with greater affinity for one of the partners in a relationship comes along it will displace the other could readily be applied to social or romantic relationships. The German polymath scientist and author Johann Goethe used Bergman's affinity table as the basis of a novella in which this happened, with social relationships subject to flux as people with greater affinity to the incumbents of a situation appear on the scene. The novella, *The Elective Affinities*, was published in 1809. Goethe openly acknowledged that it was based on Bergman's table.

From the 1770s onwards, it became apparent that the affinities varied with temperature. This meant that, logically, different tables were needed to take account of different temperatures, greatly complicating an already cumbersome system. Some other method was clearly needed for working out and recording reactivity.

'Matter may ultimately be found to be the same in essence, differing only in the arrangement of its particles or two or three simple substances may provide all the varieties of compound bodies.'

Humphry Davy, 1812

Atoms in focus

A new method would come soon, with radical changes in chemistry. In England, John Dalton turned his mind to the problem of atoms and brought clarity to two fields at once, but his work was hardly welcomed with open arms by the chemical community.

A new atomic theory

Dalton made the first attempt to explain the nature of all matter through a consistent and comprehensive theory of atoms.

There are four parts to Dalton's atomic theory. They are basically still sound and, with a few qualifications, form the foundations of all modern chemistry:

John Dalton pioneered the development of modern atomic theory.

- All matter is made of atoms, which are indivisible.
- All atoms of a given element are identical in mass and properties.
- Compounds are combinations of two or more different types of atoms.
- A chemical reaction involves the rearrangement of atoms.

It remains true that atoms are indivisible by chemical means, but they break down through radioactive decay and nuclear fission. Atoms of a given element are usually identical in mass and properties, but elements come in different forms, known as isotopes. These have the same chemical properties but a slightly different atomic mass (they differ in the number of neutrons in the atom).

Dalton described atoms as 'solid, massy, hard, impenetrable, movable particle(s)'. They remained theoretical – he had no way of demonstrating their existence. As they are indivisible, they can only combine to make compounds in ratios that are whole numbers. Water has to be written H_2O rather than $HO_{0.5}$ because half an atom is impossible (though Dalton thought it should be represented by HO).

He presented his atomic theory in 1803 in a series of lectures at the Royal Institution, but it was slow to gain credence.

Matter isn't going anywhere

Dalton based his theory on two principles: the conservation of mass, established by Lavoisier, and the law of constant composition.

Lavoisier's law states that matter cannot be created or destroyed in a chemical reaction, its constituents can only be rearranged.

JOHN DALTON (1766-1844)

Dalton began his career as a schoolmaster teaching in a Quaker school in Kendal in the English Lake District, possibly starting as early as the age of 12. His family was poor, and he was largely self-taught. He became an expert on and lectured in colour-blindness (which afflicted him) and meteorology, and it was through his meteorological studies that he became interested in the chemistry of gases. He began keeping a meteorological diary at the age of 21 and maintained it for 57 years, leaving an invaluable record for later researchers.

He derived Charles' Law independently, stating it as 'all elastic fluids expand the same quantity by heat'. He disagreed with the great Lavoisier, pointing out that the air does not act like a solvent but is a mechanical system in which everything is accomplished by the movement of particles. He brought together all his theories about gases and atoms in his *New System of Chemical Philosophy* (1808–27), though he presented them initially in lectures. In explaining his ideas about atoms, he pointed out that they must be of different sizes. He also stated that he believed elements combined in the simplest proportions – usually 1:1 – though there were no grounds for this other than an expectation of parsimony. This gave formulae such as HO (now H_2O) for water and NH (now NH_3) for ammonia.

His approach was unpopular and his ideas were treated as theoretical rather than a real description of what might be going on in matter. As a Quaker from Manchester, Dalton was looked down on by the urbane and sophisticated scientists based in London and Paris.

Fire is transformative, but the transformation effected is a reconfiguration of atoms; no matter is destroyed or created.

So while we can 'destroy' a piece of wood by burning it, all we have really done is freed the carbon, hydrogen, oxygen and other elements to reconfigure into different compounds. The total mass of the elements involved does not change. If we burn the wood in a fireplace, it will look as though mass has been lost, as the ashes have less mass than the wood. But that is the result of much of the matter having escaped in gases or water vapour.

The law of constant composition states that the same substance is always made of the same constituents. Every bit of table salt is made of sodium and chlorine, and always in the same proportions.

Comparing atoms

As we have seen, Dalton recognized that the atoms of different gases must be different sizes in order to explain the different partial pressures they exert. Having reached this conclusion, it was only natural to try to work out the relative sizes of atoms. He did this by weighing fixed volumes of gases and looking at how he thought they combined.

Lavoisier had previously reported that 87.4 parts by weight of oxygen combine with 12.6 parts of hydrogen to make water, and Dalton calculated from this that – assuming they combined in the ratio 1:1 – oxygen must be seven times as heavy as hydrogen. As hydrogen was the lightest of the elements he could find, he took this as the standard, giving it atomic mass 1. More careful measurement later revised the atomic mass of oxygen from 7 to 8. Dalton realized around 1804 that he had discovered a new and useful way of measuring elements.

He produced lists of atomic weights in 1803 and 1804 and publicized the method in 1807 and 1808. (Of course, Dalton was wrong in that he had made a fundamental mistake: hydrogen and oxygen combine in the ratio 2:1, not 1:1, and the atomic mass of oxygen is 16.)

A French chemist, Joseph Proust, developed the 'law of definite proportions' in 1794. This demonstrated that elements always combine in certain ratios by weight. He worked with the two types of tin oxide, carefully weighing the amounts of tin and oxygen needed to form each and found that one is 88.1 per cent tin and 11.9 per cent oxygen by weight and the other is 78.7 per cent tin with 21.3 per cent oxygen. The second clearly takes almost twice as much oxygen as the first. A little calculation showed that 100g of tin combines with either 13.5g or 27g of oxygen. As 13.5 and 27 form a ratio of 1:2, this supports the idea that compounds form using whole-number ratios. Proust also worked with copper carbonates and iron sulphides and found the same relationship: there are always whole-number ratios between the different weights of a substance that will combine with the metal. Dalton realized what Proust's figures meant in terms of his theory of atomic mass: 100g of tin would combine with either 13.5g or 27g of oxygen. Therefore one atom of tin could combine either with one atom of oxygen (SnO) or two atoms of oxygen (SnO_2).

A copper roof turns green over time, acquiring a patina that is a mix of three minerals.

Gases again

Gay-Lussac set about taking daring flights in hydrogen-filled balloons, measuring gas temperatures and pressures and the humidity of the air and collecting samples of the atmosphere at different levels up to 7,000m (4.5 miles) above sea level. By 1808, he had a new law that was successfully named after him: 'The ratio between the volumes of the reactant gases and the gaseous products can be expressed in simple whole numbers.' He had discovered that gases always combine in simple whole-number ratios by volume and the products are whole-number multiples of the original combining volumes. For instance, he found that two volumes of hydrogen combine with one volume of oxygen to produce two volumes of gaseous water. He couldn't explain it, but it was a solid and reproducible result.

Joseph Proust recognized that elements combine in whole-number ratios.

The Italian chemist (and count) Amedeo Avogadro (1776–1856) came up with an explanation for Gay-Lussac's finding in 1811. He suggested that equal volumes of gases at the same temperature and pressure always contain the same number of particles. So a cubic metre of oxygen would contain as many oxygen particles as a cubic metre of hydrogen, or nitrogen, or any other gas at the same temperature. Applying his ideas to Gay-Lussac's finding, combining two units of gas produces one unit of gas if the number of particles reduces.

Looking more closely, Avogadro also realized for the first time that an element might exist not as single atoms but in a molecular form. It had not occurred to anyone before that molecules didn't have to contain atoms of different elements. Deducing that both hydrogen and oxygen

EVERYTHING IS HYDROGEN

The English chemist William Prout (1785–1850) noted that the atomic weights that had then been published all seemed to be multiples of the atomic weight of hydrogen. On this basis, he suggested in 1815 that all elements are made from clumps of hydrogen atoms. He called this hydrogen building-block 'protyle'.

The hypothesis enjoyed some popularity until around 1828 when Swedish chemist Jöns Jacob Berzelius (1779–1848) published more accurate atomic weights that appeared to undermine it. In particular, chlorine has an atomic mass of about 35.5, which would necessitate making half a hydrogen atom the fundamental unit. (Some people did suggest this, but there were still discrepancies and Prout's idea fell from favour.) In 1925, the reason for chlorine's odd atomic weight was discovered – it's a mix of isotopes with atomic weight 35 and 37, and an average of 35.5.

REPRESENTING ATOMS

Dalton first set about showing the way atoms combine by using graphic symbols to represent them. He began with a different design for each element and showed compounds by putting the symbols in combinations. It didn't catch on as it put a considerable burden on printers. It's just as well, too, since the number of elements was set to rise from the 36 that he recognized to well over 100 now. In addition, while this looked fairly straightforward for molecules comprising two or three atoms, it would not serve well for more complex compounds with much larger numbers of atoms.

Fortunately, there was a better way. In 1813, Berzelius suggested the form of notation that is still in use. This involves using the initial letter of the Latin name of the element – or two letters if the first has already been assigned to another element – and then a subscript number to indicate the number of parts in combination if it is more than one: so H_2O means that two atoms of hydrogen combine with one atom of oxygen to make water. Berzelius spent most of his life trying to determine accurate atomic weights, and derived a number of formulae.

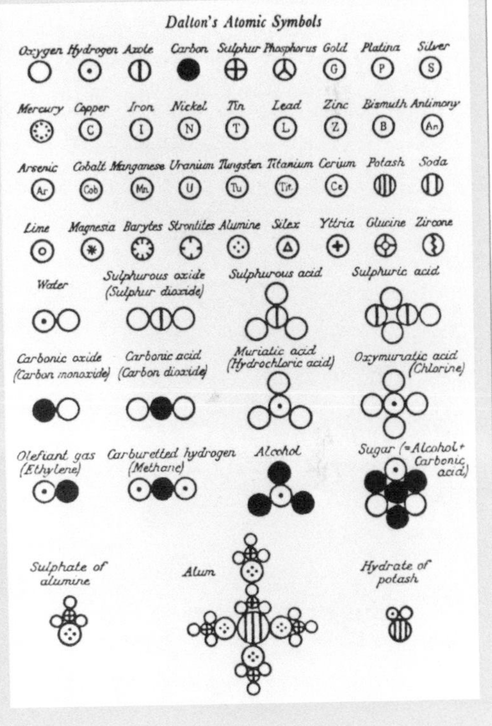

Dalton's atomic symbols and their combination to represent molecules. As the example of alum shows, the diagrams quickly become complex.

exist in molecular form, each comprising two atoms of the element, the maths of Gay-Lussac's measurement then made sense:

$$2H_2 + O_2 \rightarrow 2H_2O$$

In other words, four atoms of hydrogen and two atoms of oyxgen make two molecules of water.

From molecules to moles

Avogadro continued to work on gases, turning his attention to mass and density. It followed that if the same volume of a gas always contains the same number of particles, comparing the mass of the same volume of different gases shows the relative masses of the particles. So, if a particular volume of hydrogen weighs 1g and of

oxygen weighs 16g, we can say that the oxygen particles have 16 times the mass of the hydrogen particles.

Avogadro's findings had little immediate impact on the wider world of chemistry. He was based in Italy, which was not a centre of excellence in chemistry at the time; he published in a French journal that was not widely read; and – most importantly – he disagreed with (by then) highly regarded scientists such as Dalton and Berzelius. The full significance of Avogadro's discovery would not be understood until after his death.

Atoms by proxy

Evidence for the atom appeared in 1827 and its existence was finally proved by Albert Einstein nearly 80 years later. The English botanist Robert Brown was examining a grain of pollen under the microscope when he noticed that, instead of staying in one place, it wandered randomly around the slide. At first, he took this to be a sign of life, assuming the pollen was moving under its own volition, but then he examined a tiny object that he knew to be inanimate and observed the same kind of motion, which became known as Brownian motion.

It was explained by Einstein in 1905: water is composed of invisibly small particles (molecules) that are in constant motion. These jostle any particles they come into contact with, such as pollen grains, causing them to move. Einstein derived mathematical models of the movement of a particle under these conditions. In 1908, Jean Perrin carried out experiments with Brownian motion that

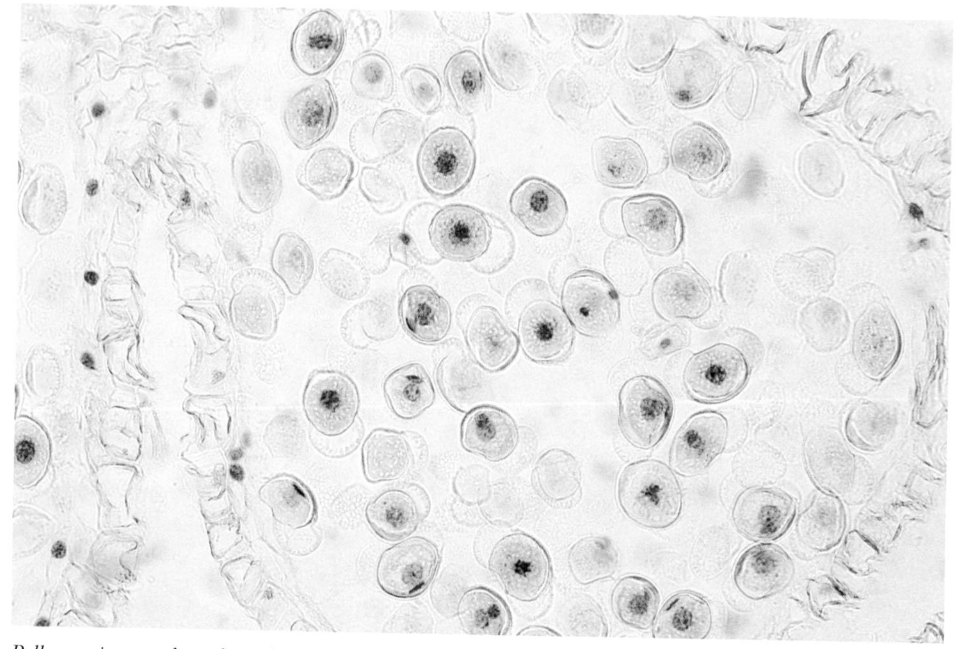

Pollen grains, seen here through a microscope, are small enough and light enough to be jostled around by molecules of water.

confirmed Einstein's predictions, proving for the first time that atoms – or at least molecules – do actually exist.

Elements get together

Although Dalton tried to work out how many atoms of elements combine to form molecules of a compound, he didn't address the mechanism of their combination. This could not be fully elucidated until the structure of the atom was understood.

There were a few early attempts at explaining how atoms are held together in compounds. Newton speculated about 'some force' that was very strong over short distances, but that is rather vague. It would be more than 100 years before something specific could be suggested.

Atomic bonds

The Irish chemist William Higgins (1763–1825) suggested something similar to Dalton's atomic theory 19 years before Dalton, though it was not as thorough as Dalton's account. Higgins was not particularly popular, and lived in Dublin, an out-of-the-way city as far as the European scientific community was concerned. His work received little notice, though Humphry Davy later helped to champion his claim for priority. But Higgins attempted something that Dalton failed to do: he explained how 'ultimate particles' (atoms) might be held together by 'force', and tried to quantify that force.

Higgins suggested that if the force between a nitrogen atom and an oxygen atom were 6 (there was no unit), then in NO_3 it would be evenly divided, with 3 coming from the nitrogen and 3 from the oxygen. If nitrogen entered into different compounds with a different number of nitrogen atoms, the 3 that was nitrogen's part of the deal would have to be divided among the associated number of atoms (see diagram, below).

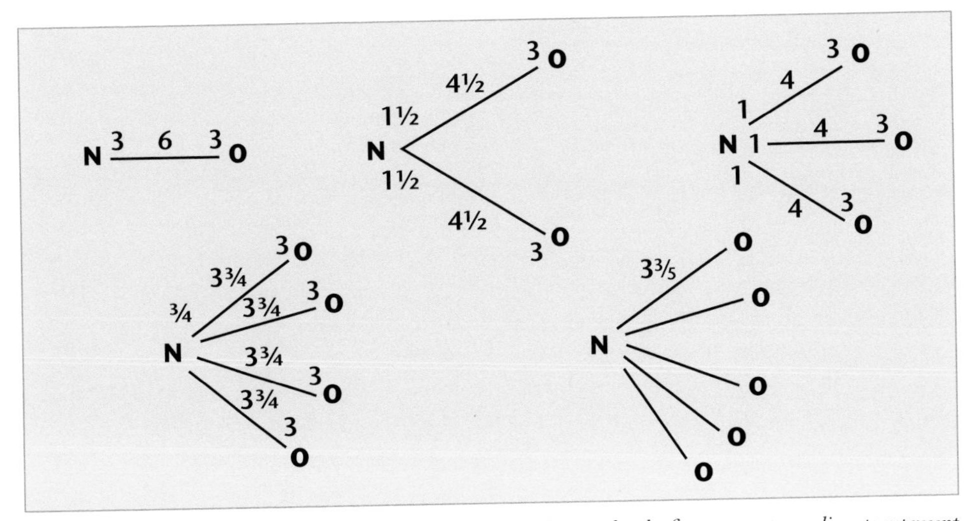

Higgins showed how 'force' might be shared between atoms. He was also the first person to use lines to represent bonds between atoms shown by the element's initial letter.

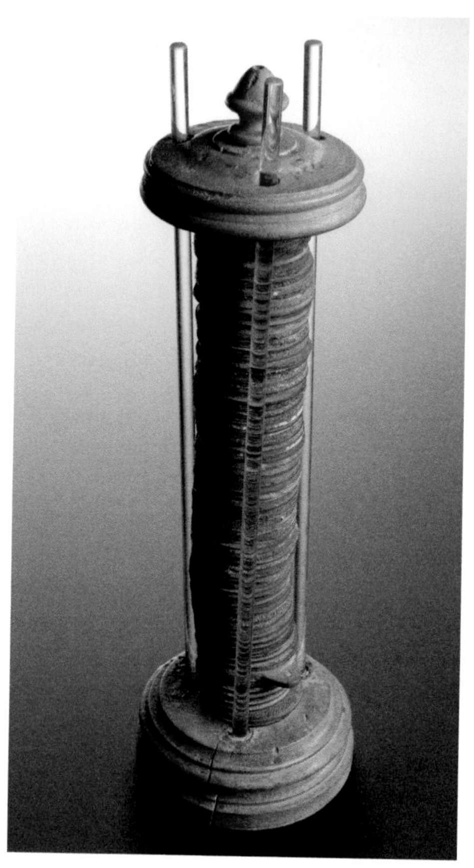

Volta's electric 'pile' – the first battery – comprised a stack of alternate copper and zinc discs interleaved with paper soaked in salt water. Connecting the the top and bottom with a wire caused an electric current to flow.

Berzelius, perhaps taking a cue from Volta's recent invention of the electric 'pile' or battery, suggested in 1819 that positive and negative electric forces might form attractions between atoms. He offered no further explanation, however, and there was no more progress until the 1850s.

Affinities become bonds

The answer to the question of how atoms form bonds began to emerge in 1852, when the English chemist Edward Frankland developed a theory to explain the affinities of different elements. He suggested atoms have a 'combining power' that allows them to link up with a certain number of other atoms. When they reach the limit of their combining power they are saturated and cannot form any more associations. He noted, for instance, that some elements tend to form compounds by combining with three atoms of another substance, such as NH_3, NO_3 and NI_3. He concluded that such elements (nitrogen in this case) are best satisfied when in such a combination.

Just a few years later, in 1858, Archibald Scott Couper introduced a new way of thinking about affinities. He thought of a bond between two atoms that held them together, and represented it as a straight line joining the symbols for the chemical elements. Just three years later, the Austrian scientist Josef Loschmidt introduced double and triple lines to represent double and triple bonds, and used a circle in a number of substances he considered probably had a cyclic structure.

A university friend of Loschmidt's, Ludwig Boltzmann, gave a definition of the chemical bond, based on the idea of regions of the atoms that are prone to form bonds:

> *'A tendency or law prevails (here), and that, no matter what the characters of the uniting atoms may be, the combining power of the attracting element, if I may be allowed the term, is always satisfied by the same number of these atoms.'*
>
> Edward Frankland, 1852

'When two atoms are situated so that their sensitive regions are in contact, or partly overlap, there will be a chemical attraction between them. We then say that they are chemically bound to each other.'

The most important work on the representation and deduction of chemical bonds and structures was carried out in organic chemistry, where the most complex molecules are found. This is the subject of the next chapter. Suffice to say here that methods of representing the structure of molecules moved ahead before the discovery of exactly how chemical bonds are formed at an atomic level.

Organizing elements

Dalton's system of explaining how atoms of elements combine to form compounds required a clear notion of which substances are elements. As we have seen, Lavoisier set out the first list, but only about two thirds of his 'elements' were actually elements.

Elements proliferate

Lavoisier listed 33 elements, of which 23 are still recognized as such. Dalton listed 36 (and they were all elements) and Berzelius identified 47. By 1850, 55 elements were known, and in 1869 the Russian chemist Dmitri Mendeleev had listed 63 elements.

Chemists were quite alarmed at the growing numbers; it wasn't so long before that people had thought there were only four. There was also a good deal of variety in their properties. Was there any way in which the emerging elements could be put into some kind of natural order?

One mole of various elements. Back row, left to right: mercury, sulphur, lead, magnesium, copper; the glass dish contains chromium; and all are standing on aluminium.

Threes, eights and spirals

The first clue that atomic weights were significant in determining the order of the elements came with the work of the German chemist Johann Döbereiner in 1817. He arranged elements in triads, or groups of three, that shared similar properties. The middle element of each triad had an atomic weight that was roughly the average of the outer two. For example, his triads included alkali-forming elements lithium (7), sodium (23) and potassium (39) and salt-forming elements chlorine (35.5), bromine (80) and iodine (127). This didn't go very far, and there were some considerable mistakes in the lists of atomic weights available at the time.

Later chemists had better luck with atomic weights. The first international chemistry congress held in Karlsruhe, Germany, in 1860 (see page 149) presented Avogadro's hypothesis. More accurate lists of atomic weights soon followed. The first person to use this revised list as the basis of a periodic pattern of elements was the French geologist Alexandre Béguyer de Chancourtois in 1862. He listed the elements in order of atomic weight on a piece of paper wrapped around a cylinder, completing a circuit at every 16th element. Unwinding the paper produced a spiral arrangement of elements that he called a 'telluric spiral' because the element tellurium came in the middle. His work

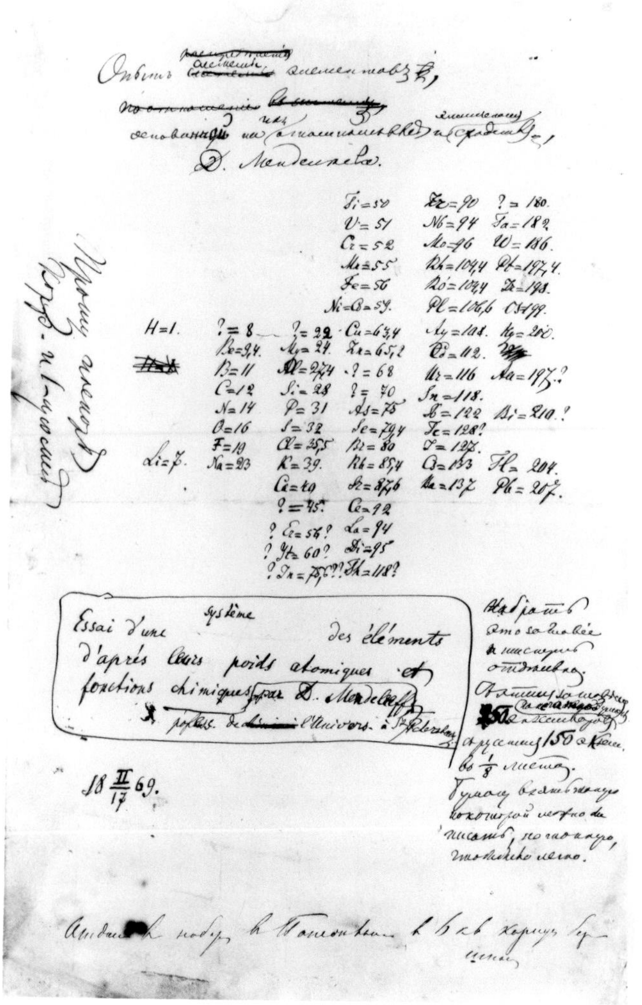

Mendeleev's first version of the Periodic Table is nothing like the table we know now. The main body is in the longer eight rows in the middle. The table has since been rotated through 90°. The mark '?' indicates elements he felt should exist, though none of that atomic weight was then known.

disappeared without making any impact, possibly because it was published without the diagram needed to make sense of it. But it wasn't as promising as it sounds. Although some elements were in the right place, others were not, or appeared twice, and some of the 'elements' he included were compounds.

When the English chemist John Newlands put the elements in order of atomic weight in 1864, he found that they fell into 'octaves', with each element having features similar to the ones eight places before or after it. His work did not attract a following, though, as it put iron in the same group as oxygen and sulphur, with which it does not share properties.

The German chemist Julius Lothar Meyer plotted the atomic weight of elements against their atomic volume (in modern terms this would be the mass of a mole of the element plotted against the volume of a mole of the solid element). He found a repeating pattern, with volume increasing and then falling against atomic weight in a periodic sequence. But it was the Russian chemist Dmitri Mendeleev who would solve the puzzle. Meyer was unfortunate in that he discovered periodicity in 1865 but did not publish until the year after Mendeleev, the originator of the modern Periodic Table. Neither was aware of the other's work.

Elements on the cards

Mendeleev, like Meyer, had been at the Karlsruhe congress and was inspired to investigate atomic weight and periodicity. He set about his task by writing the names of each of the 60 elements known at the time on to pieces of card, along with their atomic weight and characteristics. Tradition has it that while he was playing patience, it occurred to him that he could organize his cards of elements into ascending order of atomic weight and see if a pattern emerged.

> *'In a dream I saw a table where all the elements fell into place as required. Awakening, I immediately wrote it down on a piece of paper.'*
>
> Dmitri Mendeleev

DMITIRI MENDELEEV (1834–1907)

Mendeleev was born in Siberia, Russia, into a large family – he had up to 16 brothers and sisters. His father died when he was 13; his mother's glass-making factory burned down when he was 15. The young Mendeleev moved to St Petersburg and trained as a teacher. By the age of 20 he was suffering from tuberculosis and often had to work from bed. It was far from a promising start. But Mendeleev did prove to be a promising chemist – exceptionally so. He went to work with the great German chemist Robert Bunsen (see page 183) at the University of Heidelberg where he first encountered spectroscopy (see page 183). He also attended the Karlsruhe congress in 1860.

When Mendeleev returned home in 1861, his passion for chemistry was undiminished but he worried about the poor state of its teaching in Russia. He became chemistry professor of the university in St Petersburg at the age of 33, and he went on to publish two immensely successful textbooks which were also used outside Russia. His development of the Periodic Table was his crowning achievement.

Mendeleev is considered the father of the Periodic Table.

Mendeleev spent hour after hour rearranging the cards. He saw that there was some significance to putting them in order of atomic weight – similar types of properties appeared in sequence repeatedly – but he could not see the complete pattern. Eventually, he set them aside and went to sleep. The answer came to him as he slept and all he had to do when he woke up was write it down.

Mendeleev's first arrangement of the Periodic Table was rather different from the current one. His columns correspond to our rows, and his rows to our columns, but with hydrogen (H) and lithium (Li) set apart at the start. He did not know about the noble gases, so the rows go from the halogens, starting with fluorine (F) to the alkali metals, starting with sodium (Na). Looking at the bottom row of Mendeleev's table gives us the alkali metals, from lithium to caesium (Cs), though it goes awry with thallium (Tl).

In 1871, Mendeleev turned the table through 90 degrees; this is the form in which we use it.

Plugging gaps

Mendeleev wasn't the first to put the elements in order of atomic weight, but he was the most successful. The main reason for his success was that he suggested correcting some of the atomic weights that put elements in the wrong places, and left gaps for elements he felt should exist, predicting not just their existence but some of their properties. When the right atomic weights were used, the elements

fitted neatly into Mendeleev's table. Most significantly, some of the gaps that he left were filled during his lifetime when the elements he predicted were discovered.

The first predicted element to be found was gallium, which Mendeleev had called eka-aluminium. (He gave it this name because it would come after aluminium in his table, and eka is Sanskrit for '1'.) It was identified in 1875 by the French chemist Paul Emile Lecoq de Boisbaudran, who named it after the Latin name for ancient France ('Gallia'). Its properties were a close match for Mendeleev's prediction, except that de Boisbaudran gave a value of 4.9 g/cm^3 for the density whereas Mendeleev said it should be 6.0 g/cm^3. When de Boisbaudran re-checked the density, he corrected it to 5.9 g/cm^3, vindicating Mendeleev. Two more predicted elements were soon discovered: scandium in 1879, and germanium in 1886.

The validity of the Periodic Table was confirmed by these discoveries, but Mendeleev didn't live to see his final two predicted elements discovered 50 years later. He did see the discovery of the noble gases, though. At first their discovery disconcerted him, but it soon turned out that they reinforced the pattern he had identified. Element 101, discovered in 1955, was named Mendelevium in his honour.

There were problems with Mendeleev's table, even so. Some elements seemed to be transposed. Tellurium and iodine, for example, were adjacent elements. If put in the order of their atomic weights, they had properties that belonged to the neighbouring column, but if put in the column that suited their properties the atomic weights were no longer in sequence. The puzzle would not be solved until 1913, after significant discoveries about the nature of the atom.

Crystals of gallium, the first to be found of the elements that Mendeleev predicted.

Making more

The Periodic Table had a gap in column 7 for an element with atomic number 43. Between 1828 and 1908, several candidates were proposed, but all turned out to be different elements. Previous gaps had been filled, but in this case no new element came forward. Finally, in 1936, technetium filled the gap. But it was not discovered – it was made artificially. Technetium (Tc) became the first of (so far) 24 synthetic elements. In fact, it might have been manufactured as early as 1925, but the German chemists who claimed to have found it could not reproduce their result.

Technetium, like the other synthetic elements, is radioactive and unstable, which explains its scarcity as a naturally occurring element. Isotopes of technetium do occur in very small quantities on Earth, the result of spontaneous fission of uranium-238. Technetium isotopes have half-lives ranging from around 100 nanoseconds to 4.2 million years (see box, opposite). The reason no naturally occurring technetium could be found is that any that existed when Earth formed would since have decayed and changed to something else. The small amounts of technetium that do exist are the result of recent radioactive activity. It exists naturally in some red giant stars.

Further synthetic elements followed: plutonium in 1940 and curium in 1944. Once the ball had been set rolling, more were found fairly rapidly up to the current count of 24 (at 2017).

Periodic Table of the Elements

GROUP / PERIOD	1	2	3	4	5	6	7	8	9	10	11	12	13	14	15	16	17	18
1	1 H																	2 He
2	3 Li	4 Be											5 B	6 C	7 N	8 O	9 F	10 Ne
3	11 Na	12 Mg											13 Al	14 Si	15 P	16 S	17 Cl	18 Ar
4	19 K	20 Ca	21 Sc	22 Ti	23 V	24 Cr	25 Mn	26 Fe	27 Co	28 Ni	29 Cu	30 Zn	31 Ga	32 Ge	33 As	34 Se	35 Br	36 Kr
5	37 Rb	38 Sr	39 Y	40 Zr	41 Nb	42 Mo	43 Tc	44 Ru	45 Rh	46 Pd	47 Ag	48 Cd	49 In	50 Sn	51 Sb	52 Te	53 I	54 Xe
6	55 Cs	56 Ba	57-71 *	72 Hf	73 Ta	74 W	75 Re	76 Os	77 Ir	78 Pt	79 Au	80 Hg	81 Tl	82 Pb	83 Bi	84 Po	85 At	86 Rn
7	87 Fr	88 Ra	89-103 **	104 Rf	105 Db	106 Sg	107 Bh	108 Hs	109 Mt	110 Ds	111 Rg	112 Cn	113 Nh	114 Fl	115 Mc	116 Lv	117 Ts	118 Og
		*	57 La	58 Ce	59 Pr	60 Nd	61 Pm	62 Sm	63 Eu	64 Gd	65 Tb	66 Dy	67 Ho	68 Er	69 Tm	70 Yb	71 Lu	
		**	89 Ac	90 Th	91 Pa	92 U	93 Np	94 Pu	95 Am	96 Cm	97 Bk	98 Cf	99 Es	100 Fm	101 Md	102 No	103 Lr	

KEY

METALLOID

UNKNOWN CHEMICAL PROPERTIES

METAL: ALKALI METAL, ALKALINE EARTH METAL, LANTHANIDE, ACTINIDE, TRANSITION METAL, POST-TRANSITION METAL

NON-METAL: POLYATOMIC NON-METAL, DIATOMIC NON-METAL, NOBLE GAS

The current state of the Periodic Table, 2017. Attempts are being made to synthesize a hypothetical element 119 (an alkali metal, in group I).

Indivisible atoms?

Throughout the 19th century, as Dalton's theory of the atom gained traction, atoms continued to be thought of as the smallest particles, solid and distinctive as Dalton had claimed. But by the end of the century, that had to be revised as atoms seemed to be falling apart all around.

The realization that atoms could change through radioactive decay didn't sit well with the concept of the indivisible atom. The atom's name meant 'uncuttable' – its indivisibility had been its defining feature. That was about to change.

Bits and pieces

The English physicist J. J. Thomson (1856–1940) was experimenting with a cathode ray tube and a magnet in the 1880s when he found that the green beam it produced was made up of negatively charged particles only 1/2000th the weight of an atom. The only explanation he could find was that this was a sub-atomic particle, or something broken off the atom, and that Dalton was therefore wrong about the indivisibility of atoms. Thomson had discovered the electron.

Thomson devised a new model of the atom, publicized in 1897. Since matter does not have a negative charge, he concluded that there must be something balancing the negativity of the particle he had discovered, and proposed a cloud of positively charged material surrounding the negatively charged particles. His model of the atom became known as the 'plum pudding' model, with the negative electrons analogous to raisins in the positive 'pudding' of the rest of the atom. Even as Thomson was formulating his model, its nemesis was emerging.

RADIOACTIVITY AND HALF-LIFE

A radioactive substance is one that is constantly changing by emitting energy and particles (radiation). The half-life of an element is the length of time it typically takes for half of a given quantity to decay into another substance. For instance, carbon-14, used in carbon-dating, has a half-life of around 5,730 years. This means that if you had 10g of carbon-14, half of it (5g) would decay to nitrogen-14 through losing beta particles (high-energy electrons) in 5,730 years. If you were sufficiently patient, half the remainder (2.5g) would decay over the next 5,730 years, and so on. The element with the longest half-life is tellurium-128 at 7.7×10^{24} years – more than 100,000 billion times the age of the universe. No element with an atomic number higher than 99 has any use outside research, as they all have short half-lives.

From puddings to planets

In 1895, German chemist Wilhelm Röntgen discovered X-rays, and the following year the French physicist Henri Becquerel discovered radioactivity, finding that there were two parts to the rays produced by decaying uranium. This was confirmed by Ernest Rutherford (1871–1937) in 1898; he named them alpha and beta rays. Rutherford was born in New Zealand, but was working in Canada at the time. He soon found that the 'rays' are actually beams of particles. Alpha particles turned out to be helium nuclei, and beta particles to be high-energy, high-speed electrons.

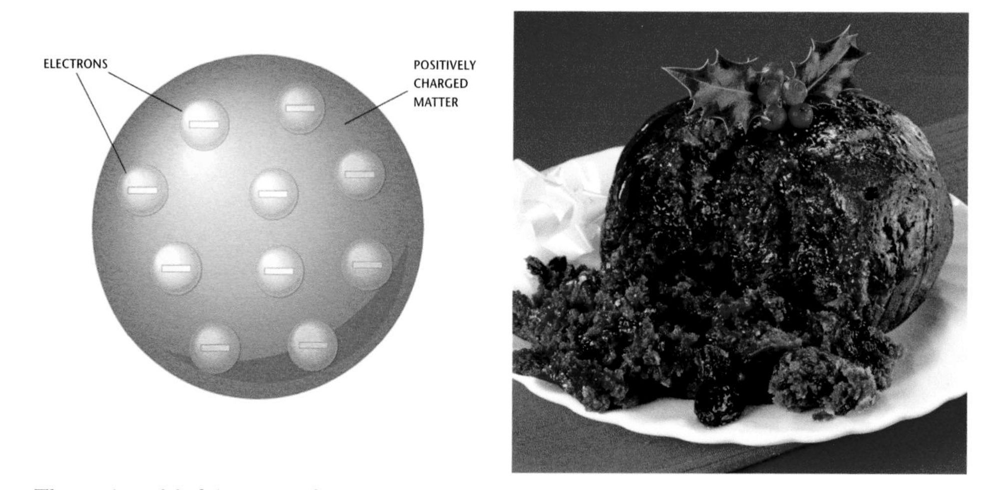

Thomson's model of the atom (left) as a cloud of positive charge (red) studded with negatively-charged electrons (yellow/green) was named the 'plum pudding' model after its resemblance to the traditional English Christmas pudding (right).

By 1907, Rutherford had moved to Manchester, England, where he worked with Hans Geiger, exploring radiation further. They fired alpha particles produced by the radioactive decay of radium through a vacuum on to a thin sheet of gold foil. Rutherford expected the particles to go straight through, perhaps with very slight deflection occasionally.

The particles produced tiny flashes of light that had to be observed and counted manually (until Geiger later invented the Geiger counter). Counting the flashes was tiresome, and in 1909 Rutherford gave the task to a research student, Ernest Marsden. He didn't expect Marsden to find anything interesting – but he couldn't have been more wrong. After a few days, Marsden found some particles deflected by very large angles and a few even bounced straight back. It was totally unexpected and completely undermined the plum-pudding model of the atom. The diffuse positive charge proposed could not repel the alpha particles to such a degree – so it must be wrong.

The only explanation Rutherford could give was that the alpha particles were being repelled by a positive charge concentrated in a small volume. The few alpha particles that directly encountered this tiny concentration of charge would be turned from their path with considerable force. Others would be deflected by a larger or smaller angle according to how closely they approached it.

Rutherford remodelled the atom to accommodate his findings. He had all the positive charge concentrated into a very small space at the centre of the atom and the negative charge scattered around it, up to a considerable distance from the centre. This explained his result: most alpha particles passed straight through, as he had originally expected, because most of the atom was

NEW RADIOACTIVE ELEMENTS

Following Becquerel's discovery of radioactivity, the Polish chemist Marie Curie (1867–1934) set out to investigate the activity of uranium. Her first finding was that the radiation from uranium comes from the element itself, not from interaction with the environment, strongly suggesting that atoms are not indivisible. Her investigations led her to discover two further radioactive elements, polonium and radium. She also noted that radium destroys tumour-producing cells – the first step towards radiotherapy for cancer. Curie became the first female Nobel Laureate in 1903, recognizing her work on radioactivity. She was awarded a further Nobel Prize in 1911 for the discovery of polonium and radium, making her the first of only four people to win two Nobel Prizes; she remains the only woman to hold two.

Marie Curie (seated), with her daughter Irène.

either a diffuse negative charge or empty space. A very small number encountered the positive nucleus and were violently repelled. The positive particles in the nucleus would later be named protons. Their existence was demonstrated by Rutherford in 1917.

Rutherford published his findings in 1911, and only two years later the Danish physicist Niels Bohr improved his model. Instead of having the electrons wandering randomly around the nucleus, Bohr proposed that they orbit in designated shells or orbitals. This has been called the planetary model, as it's similar to planets orbiting a star: each planet has to keep to its own track and can't wander off. During the 1920s, Bohr's model was refined so that the orbitals came to be considered energy levels rather than spatial locations.

From atomic weight to atomic number

The English physicist Henry Moseley solved the last part of the puzzle of the

> *'[It was] as if you fired a 15-inch shell at a piece of tissue paper and it came back and hit you.'*
>
> Ernest Rutherford

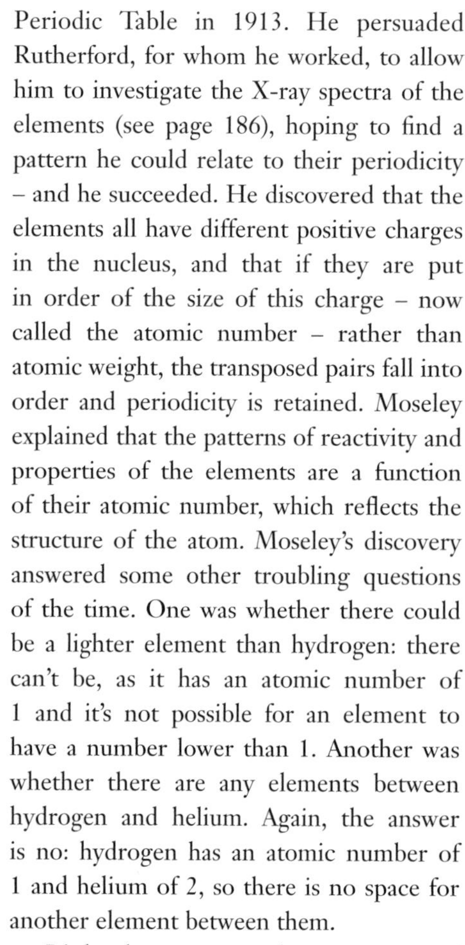

Periodic Table in 1913. He persuaded Rutherford, for whom he worked, to allow him to investigate the X-ray spectra of the elements (see page 186), hoping to find a pattern he could relate to their periodicity – and he succeeded. He discovered that the elements all have different positive charges in the nucleus, and that if they are put in order of the size of this charge – now called the atomic number – rather than atomic weight, the transposed pairs fall into order and periodicity is retained. Moseley explained that the patterns of reactivity and properties of the elements are a function of their atomic number, which reflects the structure of the atom. Moseley's discovery answered some other troubling questions of the time. One was whether there could be a lighter element than hydrogen: there can't be, as it has an atomic number of 1 and it's not possible for an element to have a number lower than 1. Another was whether there are any elements between hydrogen and helium. Again, the answer is no: hydrogen has an atomic number of 1 and helium of 2, so there is no space for another element between them.

It's hard to overstate the importance of Moseley's achievement. The atomic number is equal to the number of electrons and also the number of protons in an atom of the element (as the atom has neither positive nor negative charge). And the number of electrons, it would soon emerge, determines the reactivity of an element and how it will combine with others. But Moseley would not live to see the full impact of his discovery, made the year before the start of the First World War; he was killed during the Battle of Gallipoli in 1915.

> *'We have here a proof that there is in the atom a fundamental quantity, which increases by regular steps as we pass from one element to the next. This quantity can only be the charge on the central positive nucleus, of the existence of which we already have definite proof.'*
>
> Henry Moseley, 1913

Henry Moseley in the Balliol-Trinity laboratories, University of Oxford.

In 1932, James Chadwick discovered that the nucleus is also home to particles with no charge, called neutrons. The discovery of the neutron explained the difference between atomic mass and atomic number. The atomic mass of an element

is given by the total number of neutrons plus protons; there are generally the same number of each, so atomic mass is roughly twice the atomic number. (Electrons have negligible mass.)

Electrons put to work

As soon as it became apparent that the atom could be prised apart, at least into electrons and 'something else', a mechanism for forming bonds between atoms began to

SO – *IS* EVERYTHING HYDROGEN?

It would seem that the hypothesis suggested by Proust, that all elements are made of hydrogen, was not so far wrong after all. In fact, all elements are made from the fusion of hydrogen nuclei inside stars, but they are no longer equivalent to a collection of hydrogen atoms. When hydrogen fusion occurs, the mass of the resulting atom is not just a multiple of the masses of the hydrogen atoms involved. The 'nuclear binding energy' – energy released in the process of fusion – has to be removed. Since mass and energy are ultimately interchangeable (as demonstrated by Einstein), removing this energy reduces the mass of the system. When 56 hydrogen nuclei form an iron atom, the mass of the resulting atom is about 99.1 per cent the mass of the original nuclei, the other 0.9 per cent being lost as binding energy.

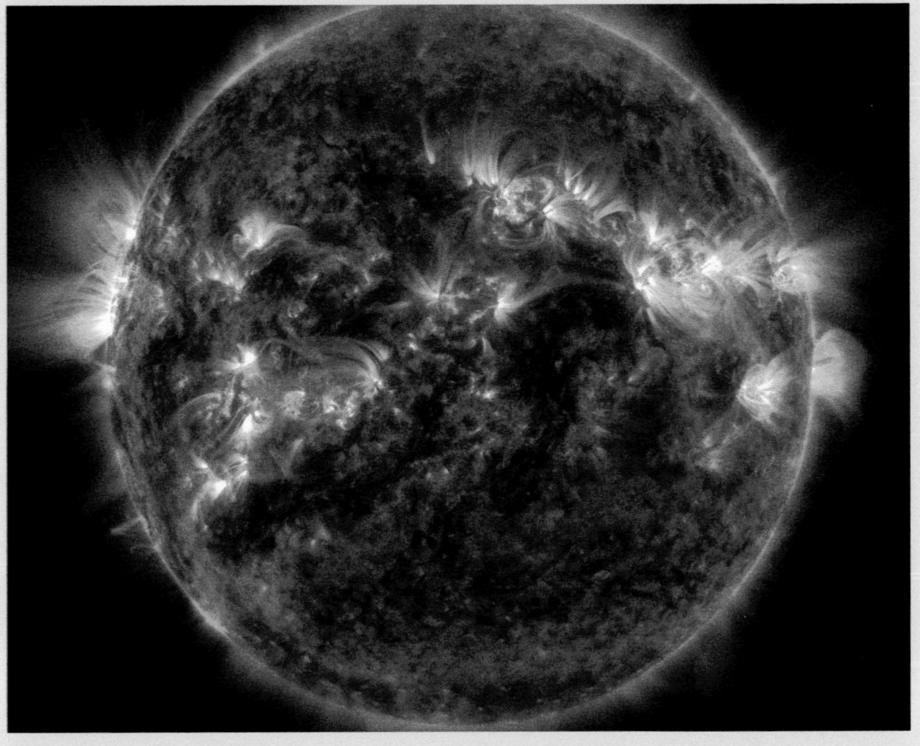

Nuclear fusion occurs inside all stars, and is the source of energy that heats our solar system.

> *'In a research which is destined to rank as one of the dozen most brilliant in conception, skillful in execution, and illuminating in results in the history of science, a young man twenty-six years old threw open the windows through which we can glimpse the sub-atomic world with a definiteness and certainty never dreamed of before. Had the European War had no other result than the snuffing out of this young life, that alone would make it one of the most hideous and most irreparable crimes in history.'*
>
> Robert Millikan, commenting on Moseley

suggest itself. The electrons, which turned out to be so far from the nucleus, could perhaps be involved in linking atoms.

Electrostatic 'tubes'

Even in the paper in which he revealed the existence of the electron, J. J. Thomson proposed that electrons might be involved in bonding. Although he did not use the word 'bond' he envisaged a 'tube of electrostatic force' that would have a positive charge at one end and a negative charge at the other end.

During the 1890s, the German physicist Wilhelm Wien was experimenting with discharge of positive ions while Thomson was experimenting with negatively charged ions. While Thomson found that negatively charged particles can exist independently of other matter (electrons), no equivalent positively charged particles could be found. The only positively charged particles around were ions. It appeared, therefore, that only negative charge could be transferred between atoms. That made electrons the obvious choice for the formation of a bond. In 1904, by which time the term 'bond' was in use, Thomson suggested bonding by means of transferring an electron wholesale between atoms: 'If we interpret the "bond" of the chemists as indicating a unit Faraday tube, connecting charged atoms in the molecule, the structural formulae of the chemist can be at once translated into the electrical theory.'

There was a problem with Thomson's thesis: it could only explain compounds formed by the complete transfer of electrons between atoms. These bonds are now known as ionic bonds. They are easily observed in electrolytic solutions, that is, liquids containing ions (particles with a negative or positive charge), with the positive and negative ions going to different electrodes.

Cubical atoms

The American physical chemist Gilbert Newton Lewis was also inspired by the notion that electrons might be at the heart of molecular bonding. Around 1902 – before Rutherford's planetary model – he proposed 'some kernel inside the atom' (see diagram, opposite) and electrons on the outside.

Lewis adopted a method of teaching his students about atomic structure that involved depicting atoms as cubes with electrons arranged at the corners. Happily, the eight corners of a cube matched the eight groups of the Periodic Table and he could draw the elements with each of their corners occupied – or not occupied – by an electron. He made a mistake in assuming that helium must have eight electrons, but apart from that, his system worked pretty well. An atom that had more than eight

> *'There seems to me to be some evidence that the charges carried by the corpuscles in the atom are large compared with those carried by the ions of an electrolyte. In the molecule of HCl, for example, I picture the components of the hydrogen atoms as held together by a great number of tubes of electrostatic force; the components of the chlorine atom are similarly held together, while only one stray tube binds the hydrogen atom to the chlorine atom.'*
>
> J. J.Thomson, 1897

electrons would begin another cube, outside the first one. He soon realized that elements would combine in such a way that they strive to fill their cubic quota of electrons, completing the octet.

Lewis continued to work on and refine his theory until he published it in 1913 and more fully in 1916. He proposed that the nature of the chemical bond is a pair of electrons, shared between atoms. For example, methane has four hydrogen atoms that each share their only electron with a carbon atom, and a carbon atom that in turn shares its four electrons with the hydrogen atoms. This gives each hydrogen atom two electrons, and the carbon atom a full complement of eight (one in each corner of his cube).

The main principles of his theory were:

- Electrons are arranged in concentric cubes in an atom.
- The number of electrons in a neutral atom (that is, one with just its own set of electrons, having neither acquired nor donated any) increases by one for each element as we step through the Periodic Table from left to right, top to bottom.
- When a cube of electrons is full, it forms the kernel of the atom around which the next octet is built up.
- If the outer cube is incomplete, the atom may give away or acquire electrons, and in doing so acquires an electric charge

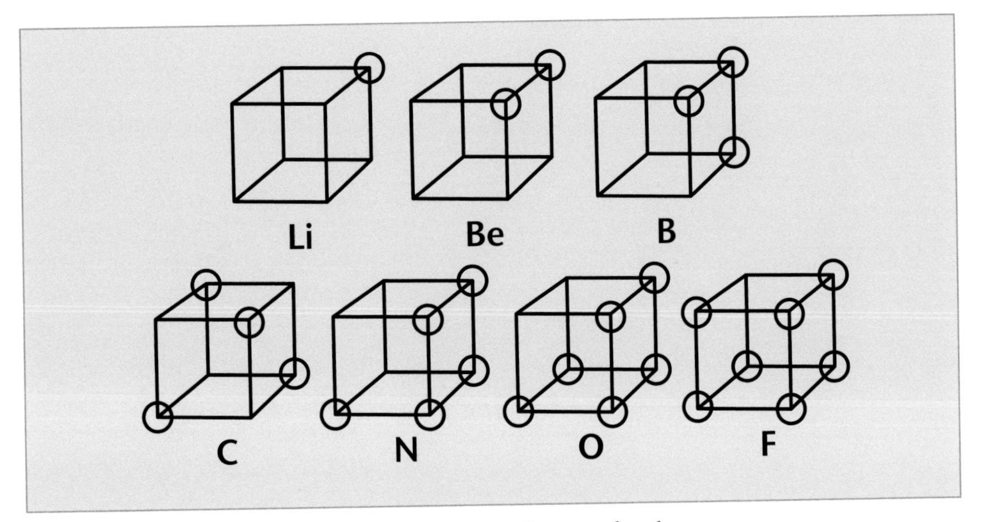

Lewis's model of the atom had a place for an electron at each corner of a cube.

> *'Two electrons thus coupled together, when lying between two atomic centers, and held jointly in the shells of the two atoms, I have considered to be the chemical bond. We thus have a concrete picture of that physical entity, that "hook and eye" which is part of the creed of the organic chemist.'*
>
> Gilbert Newton Lewis, 1923

of -1 for each electron acquired or +1 for each electron lost. This accounts for positive and negative valence (the propensity to form bonds by losing or gaining electrons).

Lewis also pioneered the use of diagrams with dots to represent electrons to show how they are shared between atoms. Although Lewis was wrong about atoms being cubes with electrons at the corners, he was right about them aiming to fill shells of eight electrons and having concentric groups of electrons around the nucleus. In 1916, he introduced the idea of paired electrons occupying the same orbital (now in Bohr's model of circular orbitals), but with opposite directions of spin.

Fuzzy and shapely

As the 20th century progressed, physicists made further advances investigating the structure of the atom. Most importantly for chemistry, the electrons' orbitals were found not to be arranged in cubes or circles around the nucleus but to have different shapes. The shape of the orbitals enables the electrons occupying them to stay as far away from each other as possible. There is an order in which the orbitals are filled, and the drive to fill octets is not as simple as it seemed, being broken into intermediate desirable states, completing pairs. Furthermore, and central to quantum theory, the location of an electron can never be fixed: the orbital simply defines a probability, an area where the electron is most likely to be found, though it could be found anywhere in the universe.

The American physicist Linus Pauling applied all the new knowledge and insights about electrons to describing the chemical bond, publishing his work first in a paper in 1931 and then in his seminal book in 1939, *The Nature of the Chemical Bond and the Structure of Molecules and Crystals: An Introduction to Modern Structural Chemistry*. Pauling avoided bogging the work down with the very complex mathematics needed to prove the point to physicists, instead making it accessible to chemists. He established six key rules about chemical bonding:

- The electron–pair bond results from the interaction of an unpaired electron on each of two atoms (each orbital can hold two electrons).
- The two electrons involved must have opposite directions of spin.
- A pair-bond is exclusive; once in a pair-bond, an electron can't form another pair-bond with a different electron.
- One wave function is involved from each atom.
- Electrons in the lowest available energy levels form the strongest bonds.
- Of two orbitals in an atom, that which can overlap the most with an orbital from another atom will form the strongest bond; the bond will tend to lie in that direction.

Linus Pauling in 1958, standing beside a model of the complex organic molecule collagen.

Pauling's findings might sound rather obscure and of no great importance; after all, if we know two atoms will combine to form a compound, do we need to know any more? But their importance lay in the provision they made for calculating the shapes and angles of bonds and molecules. This turned out to be one of the most fruitful discoveries of the 20th century. It not only allowed chemists to understand the shapes of complex molecules, but also allowed for the design of new molecules and the synthesis of chemicals whose properties can be predicted in advance.

A FINAL SPECIAL BOND

A different but vitally important bond is the hydrogen bond. First mentioned by T. S. Moore and T. F. Winmill in 1912, it is now known to be an electrostatic attraction between a hydrogen atom in covalent bond with a highly electronegative atom (such as nitrogen, oxygen or fluorine), but attracted to another highly electronegative atom nearby. In a covalent bond, the atoms share electrons. The hydrogen bond can operate within or between molecules, and is essential to life on Earth. It accounts for the high boiling point of water, as the hydrogen atoms in one water molecule are attracted to the oxygen atoms of nearby water molecules. It also enables the folding of proteins and holds the strands together by linking base pairs in DNA (see page 189).

CHAPTER 7

BONDS OF LIFE

'Organic chemistry just now is enough to drive one mad.'

Friedrich Wöhler, 1835

The early alchemists and proto-chemists worked mostly with inorganic matter – metals, salts and gases. Yet today, 98 per cent of all known compounds are organic. Organic chemistry is the branch of the science that deals with carbon compounds, originally found mainly in living things and their fossils but now also made artificially. Organic chemistry is arguably the most exciting and innovative area and certainly the most complex. Organic polymers, pharmaceuticals and biochemistry dominate modern chemistry.

Living organisms, such as vicuña and grass, work as factories producing organic chemicals.

The living and the not-living

Organic chemistry is generally considered to have started in the early 19th century. Until the end of the 18th century, living organisms were considered to have some kind of vital force, spark or spirit that distinguished them from non-living matter. This force was special, even divine, and freed organisms from the need to follow the rules of physics and chemistry which applied to inorganic matter. Consequently, it was thought that organic chemicals – those that made up living bodies – could not be synthesized but could only be extracted from an organism. This doctrine of vitalism represented a barrier to research, as chemists were unwilling to attempt something they considered impossible.

The protein in eggs is irreversibly changed by cooking.

Certain distinguishing features of organic chemicals differentiate them from inorganic substances. On the whole, they are flammable and are changed irrevocably by burning. Inorganic substances can often be heated, melted and then re-solidified with no change, but many organic substances can't be reconstituted after heating – you can't uncook an egg, for example. This seemed to support the view that there was

An alchemist in dialogue with Nature, 1516. Alchemy and the organic natural world were generally considered separate realms.

> *'Organic substances are formed by the action of peculiar organs, each organ being endowed with the power of producing different results from similar elements. . . . But although [the chemist] can decompose the products of organic action, and find the proportions of their elements, he never has been able to recompose or imitate these compounds.'*
>
> J. L. Comstock, *Elements of Chemistry*, 1835

ORGANIC AND INORGANIC

The distinction between organic and inorganic chemistry was first made by Berzelius in 1806. He defined organic compounds as those produced by and making up living organisms, and inorganic compounds as those found in non-living matter. The definition has since been refined to call 'organic' those compounds that contain carbon, whether or not they have a biological origin or context. But it is the German chemist Justus von Liebig (1803–73) who is considered the father of organic chemistry. He argued against vitalism or any special status for organic compounds, and spent many years isolating and exploring organic compounds, studying how they could degrade or change into others, and trying to discover their roles in living organisms.

something very different and distinctive about organic material.

There are three main sources of organic compounds today: living organisms; organic fossilized deposits made from previously living organisms, such as coal and oil; and human manufacturing processes. Organic compounds don't occur naturally in large quantities outside living organisms. A few carbon compounds, such as carbon dioxide, are considered inorganic.

FROM CARBON TO CUBES

Von Liebig had a large family and never earned a great deal from chemistry. As a consequence, he was always on the look-out for other ventures that might turn a profit. He was also concerned about the nutrition of the poor. In 1847, he developed a method of producing a cheap meat extract so that everyone could enjoy the benefits of meat. It was too expensive to produce in Europe, where meat was costly, but in 1862 he joined with a partner who suggested making it in South America where cattle were reared for leather. With no way to preserve and transport it, most of the meat was left to rot. The product eventually became the Oxo cube, used as gravy and for flavouring stews and casseroles.

Breaking down, not building up

As it was believed that organic compounds could not be made outside a living organism, chemists focused on trying to discover their components rather than synthesizing them. Lavoisier was the first to develop methods of analysis which could determine the quantities of carbon and hydrogen in organic material and reveal its empirical formula. He did this by burning the materials in oxygen and weighing the quantities of water and carbon dioxide produced. This revealed little other than the comparative proportions of carbon and hydrogen, and unless a sample of a simple, pure organic compound was the starting point it revealed nothing very useful at all. Inevitably, too, some carbonized remains were left over so the method was somewhat imprecise.

The process was steadily improved by burning material in the presence of oxidizing agents. The French chemist Jean-

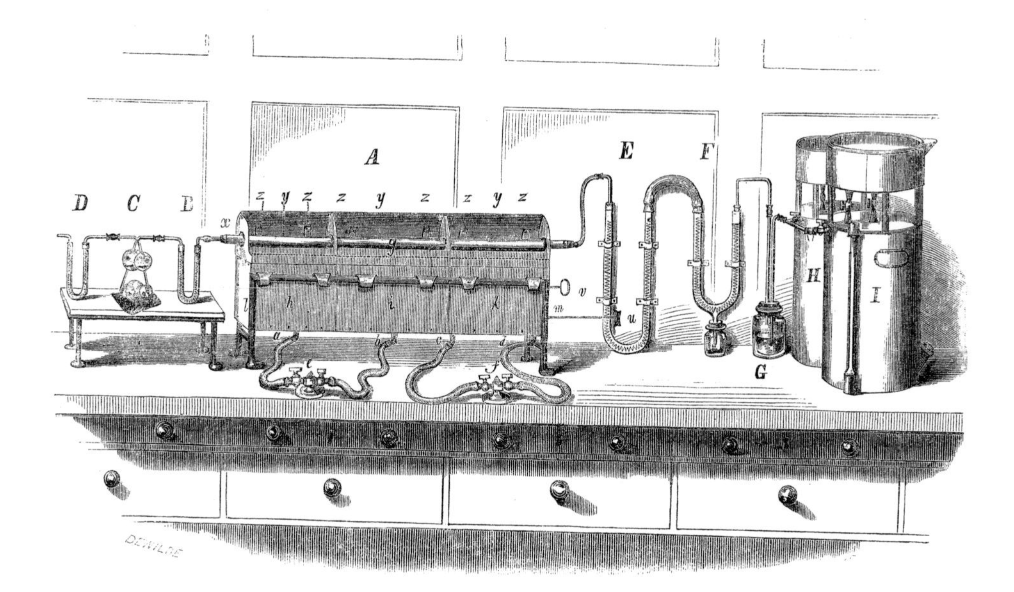

Liebig's apparatus for organic analysis, adapted to use gas as a source of heat.

Baptiste Dumas learned how to determine nitrogen, and von Liebig found a way to measure sulphur and the halogens – but still the results gave only ratios and no information about structure, which is vital in organic chemistry.

Complex organic substances such as starch, fats and proteins presented a considerable problem to the chemists who wished to analyze them. Some, it turned out, could be broken down into building blocks by treatment with dilute acid or alkali. In 1812, the Russian chemist Gottlieb Sigismund Kirchhoff (1764–1833) heated starch with acid and reduced it to a single, simple sugar eventually named glucose. (Starch is made up of many glucose units joined together.) In 1820, the French chemist Henri Braconnot used a similar process to isolate the first amino acid, glycine, from the protein gelatin. Amino acids are the building blocks of all proteins.

Vitalism – a load of pee?

In 1773, the French chemist Hilaire Rouelle issued the first chemical challenge to the theory of vitalism. He managed to derive crystals of urea from the urine of several animals, including human beings, and found it to have a relatively simple make-up. This contradicted the accepted view that all organic compounds were highly complex and could not be made in the laboratory – either with or without a vital spark. But the revelation by the German chemist Friedrich Wöhler (1800–82) that he could actually synthesize urea without recourse to a kidney was a fatal blow to vitalism. He was distraught by the implications of his discovery.

> *'The great tragedy of science, the slaying of a beautiful hypothesis by an ugly fact.'*
> Friedrich Wöhler, 1828

In 1828, Wöhler had not set out to test vitalism or to make urea. He had been trying to make ammonium cyanate by mixing ammonium chloride with silver cyanate using this reaction:

$$AgNCO + NH_4Cl \rightarrow NH_4NCO + AgCl$$

However, his experiment yielded some strange crystals that were not ammonium cyanate and looked suspiciously like the urea crystals Rouelle had discovered. Further investigation proved they were just that. Ammonium cyanate had formed as he expected, but is very unstable. The molecules rearranged themselves spontaneously to form urea, which has the same atoms but combined in a different way:

$$NH_4NCO \rightarrow H_2N\text{-}CO\text{-}NH_2$$

Having produced an organic compound from inorganic ingredients, Wöhler had to conclude that there was, at least in urea, no vital force or spark of life.

Not only, but also

As well as undermining vitalism, this discovery confirmed that two different compounds could have the same empirical formula. Wöhler had already discovered that the silver cyanate he was using had

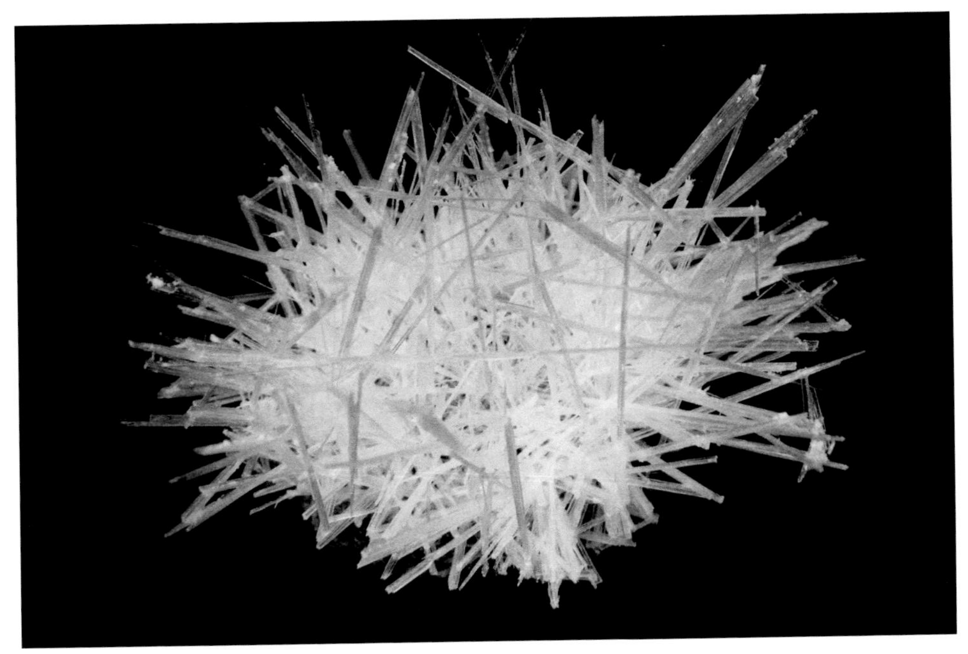

Needle-sharp crystals of urea. Equally sharp crystals of sodium urate forming in the joints cause the fierce pain of gout.

the same composition as the silver fulminate von Leibig had produced the previous year, but with different properties. Now, with urea and ammonium cyanate, he had a second example of the phenomenon. Berzelius introduced the term 'isomer' to denote such pairs of compounds in 1830. He was the first to suggest that the properties of substances are determined not only by the number and type of the constituent atoms, but also by how those atoms are arranged. Isomerism would become hugely important, not only in the laboratory but also in explaining the mystery of chemical bonding.

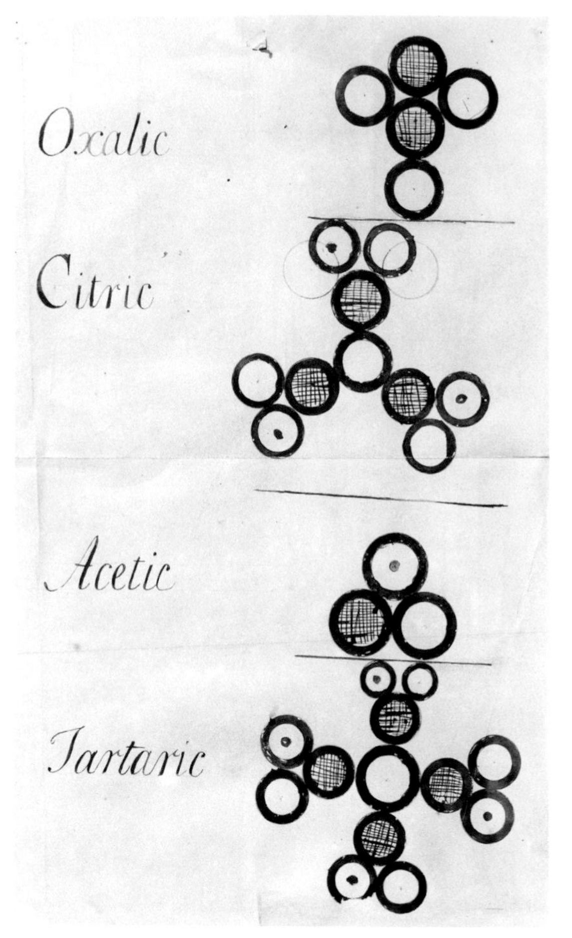

Dalton's formulae for four organic acids, drawn using his own system of symbols for each atom.

A step ahead

Even before Wöhler accidentally made urea, there had been steps towards synthesizing organic compounds. The French chemist Michel Chevreul worked with animal fats and fatty acids in 1816. Having separated the different acids, he discovered that he could make chemical changes to them, so effectively creating a new organic compound without any 'vital spirit'. This in itself was not catastrophic to vitalism, though. Chevreul was starting with an organic compound produced in the usual way – that is, by a living thing – and then making a little change to it. He was not making an organic compound from scratch.

The end of living factories

Until 1828, there had been no means of producing organic compounds besides co-opting a living organism. A chemist who wanted urea had to get it from urine. A chemist who wanted acetic acid had to get it from vinegar (derived ultimately from plant juices). Even after Wöhler's discovery, there was no great rush of people synthesizing urea. But that changed in 1845, when Hermann Kolbe showed that acetic acid could be made from carbon disulphide.

THE STORY OF MAUVE

Mauve fabrics became very fashionable in the late 19th century, and it wasn't just a quirk of transient taste. Mauve was an entirely new colour: there had been no way of dyeing fabric mauve before an 18-year-old student of chemistry, William Perkin, had an accident while trying to make quinine.

Quinine has been used to treat malaria in the West since at least 1632, when Spanish conquerors learned of its properties from indigenous South Americans. It is found in the bark of the cinchona tree and was first isolated in 1820. As cinchona trees grow a long way from Europe, synthetic quinine was desirable – especially as malaria was a serious threat to European colonial ambitions in tropical countries. Making quinine was set as a challenge to Perkin by his professor.

One of the methods Perkin tried seemed to fail, producing a nasty-looking black lump. But the lump turned out to produce a beautiful shade of purple, a dye that became known as Perkin mauve, or mauveine. Perkin started a factory to make the dye that was so commercially successful he became quite rich. At the same time, organic chemistry had a tremendous boost in terms of public visibility and popularity. Perkin's success was relatively short-lived, as mauveine spawned an industry in synthetic dyes that soon led to it being superseded. The composition of mauveine proved difficult to work out, and its molecular structure was only uncovered in 1994.

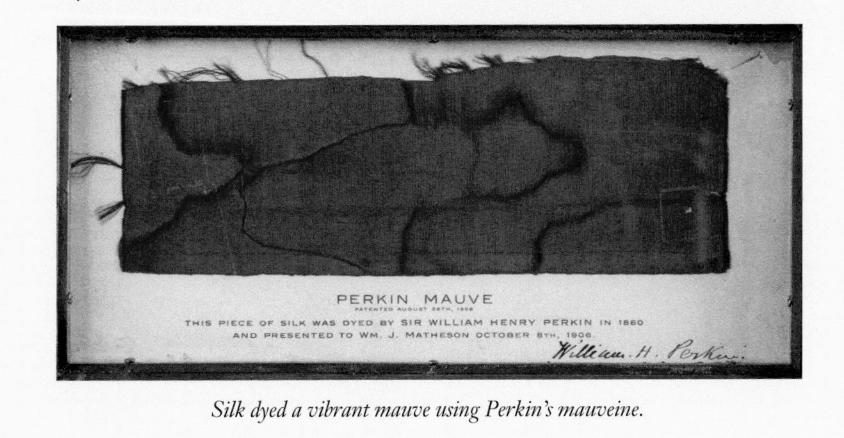

Silk dyed a vibrant mauve using Perkin's mauveine.

This was even more damaging to vitalism than Wöhler's success, as Kolbe began with inorganic compounds. Wöhler had started with two organic compounds, so it could be argued that the vital force was already present. In Kolbe's acetic acid, there was no source for the vital force. As if that wasn't enough, in 1860 Marcellin Bertholet launched a vigorous campaign against vitalism, maintaining that all chemical phenomena rely on measurable physical forces and that there is nothing remotely mysterious or special about them. He successfully produced many organic compounds, including hydrocarbons, fats and sugars, directly from inorganic ingredients. Vitalism was well and truly dead.

Out of bonds

The realization that there could be different isomers with the same chemical formula led chemists to think more about chemical bonds.

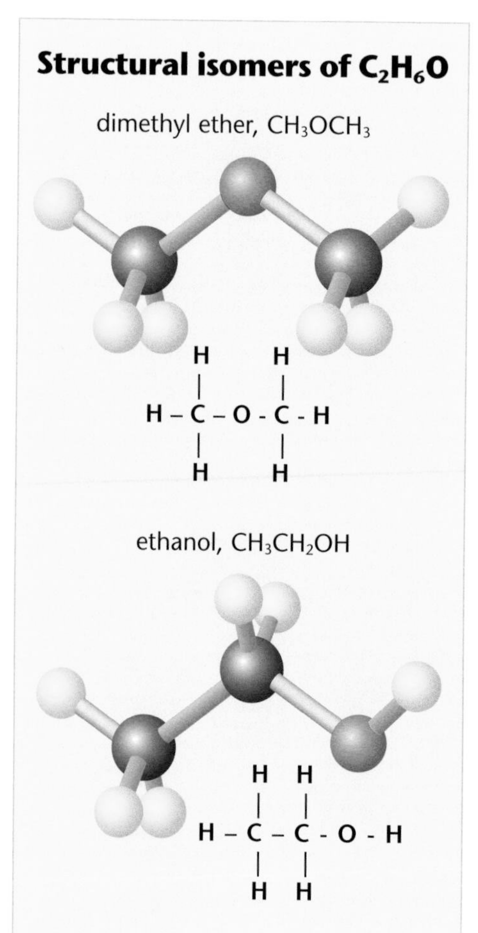

The two compounds ethanol and dimethyl ether both contain the same atoms – two carbon, six hydrogen and one oxygen – but the atoms are arranged differently, producing compounds with different characteristics. Ethanol is an alcohol, liquid at room temperature, and dimethyl ether is an ether, a gas at room temperature.

Carbon is special

Carbon forms a greater variety of compounds than any other element. The reason it can do this – and the reason it makes an excellent basis for living organisms – is that carbon atoms can join together to form long chains. The chains can have branches, and chains or branched networks can also be folded into different shapes. This gives

ALIEN BIOLOGY

Some science fiction writers imagine an alien biology based around the element silicon instead of carbon. In theory, this is entirely possible. Silicon, like carbon, has four unpaired electrons and is relatively abundant. There are problems though: silicon is a larger atom, it does not readily form double bonds and forms compounds with fewer other elements than does carbon. On Earth, carbon dioxide is a gas that plays an essential role in living processes. Silicon dioxide is a solid – silica – the main constituent of sand. For silicon to form the basis of alien life forms, the chemistry of their world would be very different from ours – and very inhospitable to us.

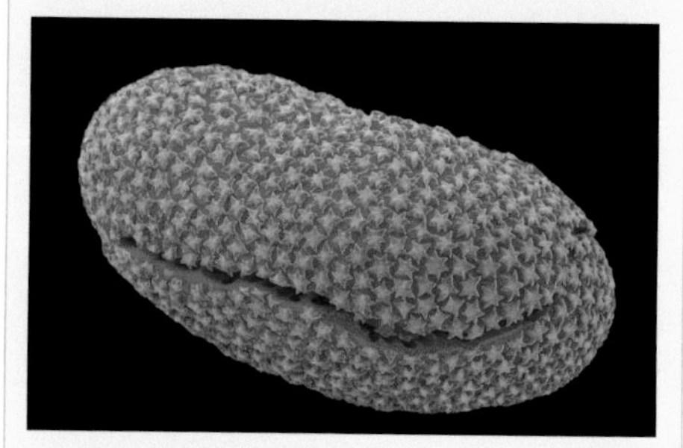

A false-colour image of a marine diatom, a carbon-based life form which extracts silicon from seawater and uses it in building its cell walls.

the potential for an infinite number of different arrangements of carbon with other elements. The elements most commonly found in combination with carbon in organic compounds are hydrogen, oxygen and nitrogen.

Carbon's ability to form myriad complex shapes comes from its valency (its ability to combine with other elements to form compounds). Discovering the details of carbon bonds and their variety was the key to understanding organic chemistry and putting it to use.

Carbon chains

The German chemist August Kekulé (1829–96) formulated the theory of chemical structure in the 1850s. In 1857, he announced the tetravalency of carbon – the ability of a carbon atom to form four bonds – and the following year he published on the ability of carbon atoms to form long chains. The same year, the Scots chemist Archibald Couper independently discovered the ability of carbon atoms to link together; he also introduced the method of showing molecular structures using lines to represent bonds between atoms. Kekulé built up chemical structures using *Verwandtschaftseinheiten* (affinity units) to show how atoms connected to one another in a coherent structure, and in doing so made both analysis and synthesis of organic compounds much easier. Organic chemists became much more productive as a consequence.

Chemistry in crisis: Karlsruhe

It might seem as though everything was going well for organic chemistry, but there was a major problem. When Kekulé brought out his book on organic chemistry, he gave 19 different formulae that had been suggested for acetic acid. Acetic acid is not even a complex organic compound; its correct formula is CH_3COOH. The problem arose because atomic weights could not be determined from combining weights of elements without certain knowledge of the formulae of some basic, simple compounds. Consequently, no other formulae could be derived with any confidence or accuracy.

Many chemists tried to work with combining weights, but it led to problems. Combining 16g of oxygen with 2g of hydrogen produces 18g of water. This gives equivalent (combining) weights of 8 for oxygen and 1 for hydrogen, yet the

Kekulé revolutionized organic chemistry by starting from molecular structure.

atomic weights are 16 for oxygen and 1 for hydrogen. We can only deduce the atomic weight if we know that the formula of water is H_2O. Charles Gerhardt suggested the system of using atomic weights, but few people accepted it. There seemed no way forward.

So, in 1859, Kekulé proposed an international symposium to tackle and discuss the problems facing chemistry. It was held in Karlsruhe, Germany, in 1860. The invitation, sent to all leading European chemists, called for a meeting to formulate: 'More precise definitions of the concepts of atom, molecule, equivalent, atomicity, alkalinity, etc.; discussion of the true equivalents of bodies and their formulas; initiation of a plan for rational nomenclature.'

In that the conference did not reach an agreement, it could be considered a failure. But it brought the matter into the light and defined the problem. The most important contribution was made by the Italian chemist Stanislao Cannizzaro (1826–1910) on the final day. He distributed a pamphlet that gave historical background to the difficulty over atomic weights and argued the case for hydrogen gas being considered as H_2. His explanation of how atomic weights should be deduced from the lowest combining weight ever observed was persuasive and after the conference Gerhardt's system became mainstream. Cannizzaro used Avogadro's finding that the mass of a fixed volume at a fixed temperature always contains the same number of particles, so the relative molecular mass of a gas can be calculated from the mass of a sample of known volume. The molecular mass can be a whole-number multiple of the atomic mass: for example, in the case of hydrogen, which exists as a diatomic molecule (that is, two hydrogen atoms are bound together), the molecular mass is twice the atomic mass.

This was the point, four years after his death, when the importance of Avogadro's work was recognized. Barely acknowledged in his lifetime, Avogadro is now considered one of the major figures in the development of molecular chemistry. In 1894, Wilhelm Ostwald named the number of particles in 1g of hydrogen or 16g of oxygen a 'mole' (from the German *Molekül*, 'molecule'), and it was finally calculated by Jean Perrin in 1909 at $6.022140857(74) \times 10^{23}$. The figure emerges logically from Avogadro's work and was named in his honour.

Dreaming of snakes

Even with the notions of chains and cross-links, not every molecular structure could be clearly fitted to the model of tetravalent carbon. One that puzzled Kekulé and others was benzene. Its empirical formula is C_6H_6; there seemed to be no way of making this work, with each hydrogen atom allowed only one bond and each carbon atom allowed four. How the solution came to Kekulé has become legendary. After pondering the problem fruitlessly, he claimed he was daydreaming when he saw an image of a snake eating its own tail – the ancient ouroboros. He realized that if he put the carbon atoms into a ring, he could make the formula of benzene work. So he arranged the carbon atoms in a hexagon, with alternate double and single bonds between them, using up three of their four potential bonds; each could then also bond

The ancient alchemical symbol ouroboros (above left) inspired Kekulé to arrange the six carbon atoms of benzene in a ring.

with a single hydrogen atom. He published his solution in 1865. In response to criticism relating to isomers (see below), he adjusted his model in 1872 to have the single and double bonds constantly switching places, so that each was single half the time and double half the time, making all bonds equivalent. For this reason, benzene is often represented as a hexagon enclosing a ring, rather than with alternate single and double bonds.

Physical evidence to support Kekulé's structure came from his investigation of benzene derivatives. He found that when a single extra chemical or group replaces one of the hydrogen atoms (a monoderivative) there is only one version, not two, so all carbon bonds are equivalent. And that when two extra elements or groups are added, replacing two hydrogen atoms (a 'diderivative'), three isomers can be produced. He explained this as resulting from the number of carbon–carbon bonds lying between the replaced hydrogen atoms: one, two or three. The number of isomers possible suggests all carbon bonds are equivalent. Derivatives of benzene – all compounds that include one or more benzene rings – are known as aromatics. (Many, but not all, aromatic compounds are aromatic in the normal sense – that is, they have a smell.)

Up and down

The common way of representing molecular structures, with bonds shown as lines between atoms, suggests that a molecule lies entirely in one plane. The recognition that this is not so and that molecules occupy three-dimensional space is at the heart of 'stereochemistry'.

Mirrored molecules

In a simple inorganic compound, there is usually only one way of arranging the atoms. In a larger molecule, and particularly in organic compounds, there can be several ways of arranging the same atoms, as we have seen, producing isomers with the same empirical formula. But an even

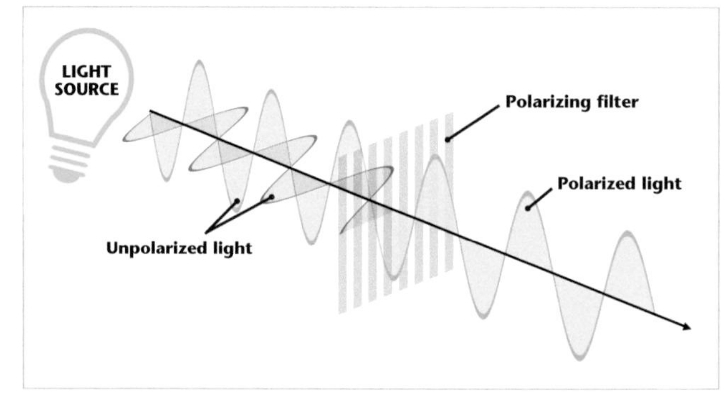

Light usually vibrates in multiple directions, but polarized light vibrates in a single plane. The action of a chiral crystal such as tartaric acid is to rotate the plane in which polarized light vibrates.

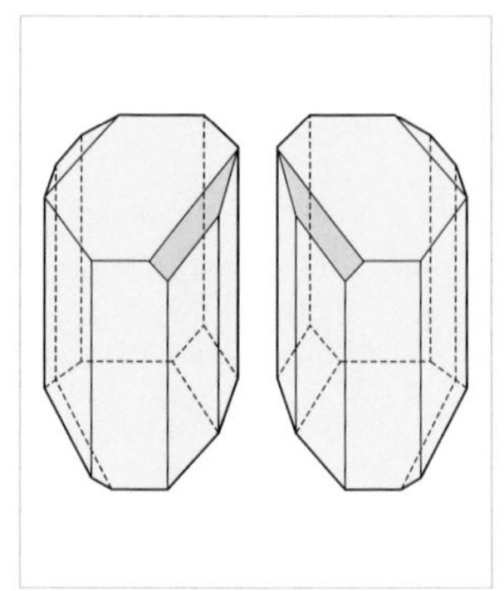

The optical isomers of tartaric acid are mirror images of each other.

more subtle difference is that the same arrangement can be differently oriented, so that in two versions each atom has the same position, but one molecule is a mirror image of the other. This is called an optical isomer or enantiomer.

Between 1815 and 1835, the French physicist Jean-Baptiste Biot discovered that several organic compounds in liquid state or in solution rotate polarized light. These include turpentine, sucrose, camphor and tartaric acid. He deduced that there was

CHIRALITY

Chirality is effectively the 'handedness' of molecules. Chiral objects cannot be superimposed on their mirror images, just as a left hand can't be put over a right hand and cover it exactly if both are the same way up.

The chirality of a molecule can completely change how it interacts with other chemicals and sometimes with the body. The drug thalidomide, given to pregnant women as a treatment for nausea in the 1960s, caused serious birth defects. It was later found that only one of the two optical isomers of thalidomide causes deformity; the other works effectively as a sedative. Even if the correct isomer had been used, though, the molecule is capable of converting between forms in the body so the defects would not have been avoided by more careful production of just one isomer of thalidomide.

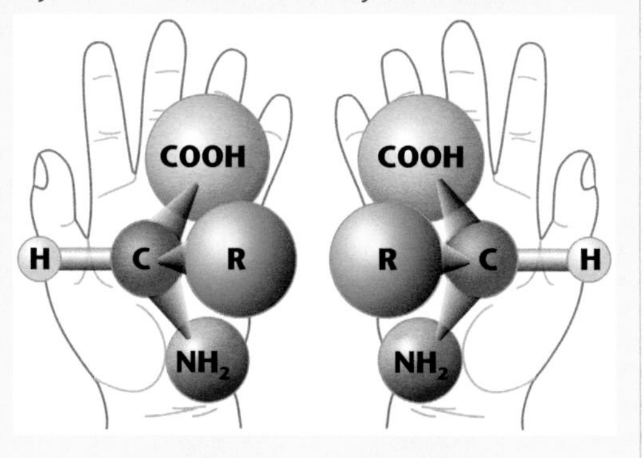

Optical isomers of a generic amino acid; R represents a side chain that is different for each amino acid.

something in the structure of the molecules that had this effect. In 1820, a by-product of the production of tartaric acid turned out to be chemically identical to tartaric acid but did not rotate polarized light – a phenomenon that could not be explained.

The puzzle was solved by the brilliant French chemist and biologist Louis Pasteur (1822–95) when he was only 25 years old. Comparing samples of both tartaric acid and the new acid, named racemic acid by Gay-Lussac in 1826, he found that although both tartaric and racemic acids had crystals of the same shape, there were two varieties of crystals in racemic acid. One had particular facets facing left and one had the same facets facing right. Separating the two, he found that both rotated polarized light but in opposite directions. In the mix of racemic acid, they cancelled each other out. He realized that the crystals, and indeed the molecules, were mirror images of each other. Lord Kelvin would call them 'chiral' in 1893 (from the Greek word for hand, to suggest left-handed and right-handed). It soon became clear that the chirality of an organic molecule affects how it interacts with a living body. Chirality is extremely important in the way that food and drugs are absorbed and work.

Asymmetric carbon

The physical nature of chiral molecules was explained by the Dutch chemist Jacobus van 't Hoff (1852–1911) and the French chemist Joseph Le Bel independently and in the same year, 1874. Van 't Hoff is considered the father of physical chemistry, and won the first Nobel Prize for Chemistry in 1901.

Even before publishing his doctoral thesis, van 't Hoff produced a pamphlet in which he proposed a tetrahedral model for the carbon atom. If each of the carbon bonds is attached to a different atom or group, the carbon atom is asymmetric. By changing the arrangement of the bonds, different isomers can be constructed from the same ingredients. He described the carbon atom as tetrahedral, meaning not that it actually has that shape but that the carbon atom is a sphere in the middle of a tetrahedron defined by the bonds it will form. Bonds coming out in different directions explained the different properties of isomers, which previously looked indistinguishable from their formulae.

The drug thalidomide caused birth defects, typically shortened or missing limbs, in the children born to some women who took the drug during pregnancy.

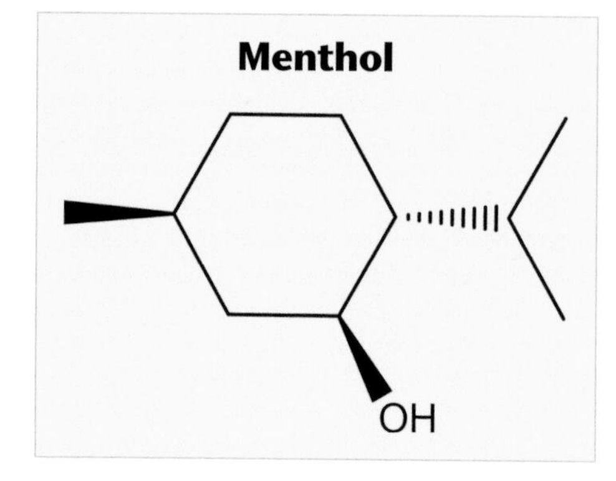

Menthol represented using van 't Hoff's convention. The benzene ring lies in the plane of the page; the OH group is in front of the page; the bond (to two CH_3 groups) on the right recedes into the page.

The idea initially attracted ridicule and vicious criticism. The German chemist Hermann Kolbe denounced van 't Hoff for having 'no liking for exact chemical investigation' and instead launching into a flight of fancy about how atoms arrange themselves in space. By 1880, though, the idea was beginning to gain ground.

Van 't Hoff developed a way of representing molecules in three-dimensional space. A bond that comes out of the plane of the page is represented by a wedge-shape – the fat end of the wedge sticks out of the page. A bond represented by a dotted line recedes into the page. Simple lines represent bonds on the same plane as the page.

The chemistry of life

From the middle of the 19th century, organic chemistry flourished. Vitalism had been debunked, and the bonds between atoms in complex compounds were becoming clear. Chemists discovered the formulae and structures of many naturally occurring organic compounds and began to synthesize them in the laboratory. They also began to make, both deliberately and accidentally, organic compounds which do not occur in nature.

At the same time, the field of biochemistry began to emerge. Once it was recognized that the chemistry which takes place in living organisms is fundamentally the same as any other chemistry, the processes of living organisms were up for investigation. The equivalence of chemical

MECHANICS OR CHEMISTRY?

In the 17th century, physiologists disagreed about whether digestion was a chemical or mechanical process. Did acid break down food in the gut, or was food broken up by the teeth and then the churning of the gut? Some quite unsavoury experiments suggested chemicals were at work. In the 18th century, the French scientist René de Réaumur experimented with hawks and Lazzaro Spallanzani carried out investigations with a variety of animals, and even himself, using vomited food and gastric juices to try to emulate digestion outside the body.

The question was settled conclusively in the 1820s when the American surgeon William Beaumont had the opportunity to experiment with digestion *in situ*. Experimenting on a patient who had a hole through his abdomen into his stomach (as a result of a gunshot wound), he found that food dissolves in the same way in the gut or in gastric juices extracted from the gut and heated. This proved that digestion was definitely a chemical process.

reactions in living and non-living contexts would be conclusively demonstrated in 1896 by Eduard Buchner.

Chemistry in and out of bodies

Humans have taken advantage of the actions of yeast in fermenting sugar to make alcohol for at least 9,000 years; the earliest evidence of an alcoholic drink (made from honey, rice and fruit) dates from Neolithic China. But no one understood the process behind it until the 19th century. Then, in the 1850s, Louis Pasteur investigated fermentation and the action of bacteria in spoiling wine, milk and other foodstuffs. He found that yeast, a living organism, is necessary to the process of fermentation, announcing his result in 1856: 'Alcoholic fermentation is an act correlated with the life and organization of the yeast cells, not with the death or putrefaction of the cells.'

A flask of the type used by Louis Pasteur in his fermentation experiments. He found fermentation only occurred if microorganisms from the air (yeast) could enter the liquid in the flask.

In 1896, Buchner proved that Pasteur was not quite right, demonstrating that the function performed by yeast continues even if the yeast is destroyed and only the contents of its cells are present. He even isolated the enzyme responsible, which he called zymase. Although produced by the yeast, it is just a chemical and performs the same function whether in a living cell or isolated in a test-tube. Chemistry, it seemed, is chemistry, no matter where or how it occurs.

Enzymes are catalysts, facilitating or speeding up chemical reactions. In 1926, the American chemist James Sumner showed that the enzyme urea is a pure protein and can be crystallized. This was contentious as people did not believe at the time that proteins could be catalysts. John Northrop and Wendell Stanley demonstrated definitively in 1929 that three digestive enzymes are, indeed, proteins. This opened the door to discovering the molecular composition of enzymes by X-ray crystallography (see page 186). The first enzyme to have its structure elucidated was lysozyme, found in tears, saliva and egg white. It was revealed by the English chemist David Phillips in 1965.

Chemical cycles

Once it became clear that the processes going on in living bodies are chemical, other questions arose. Which chemicals

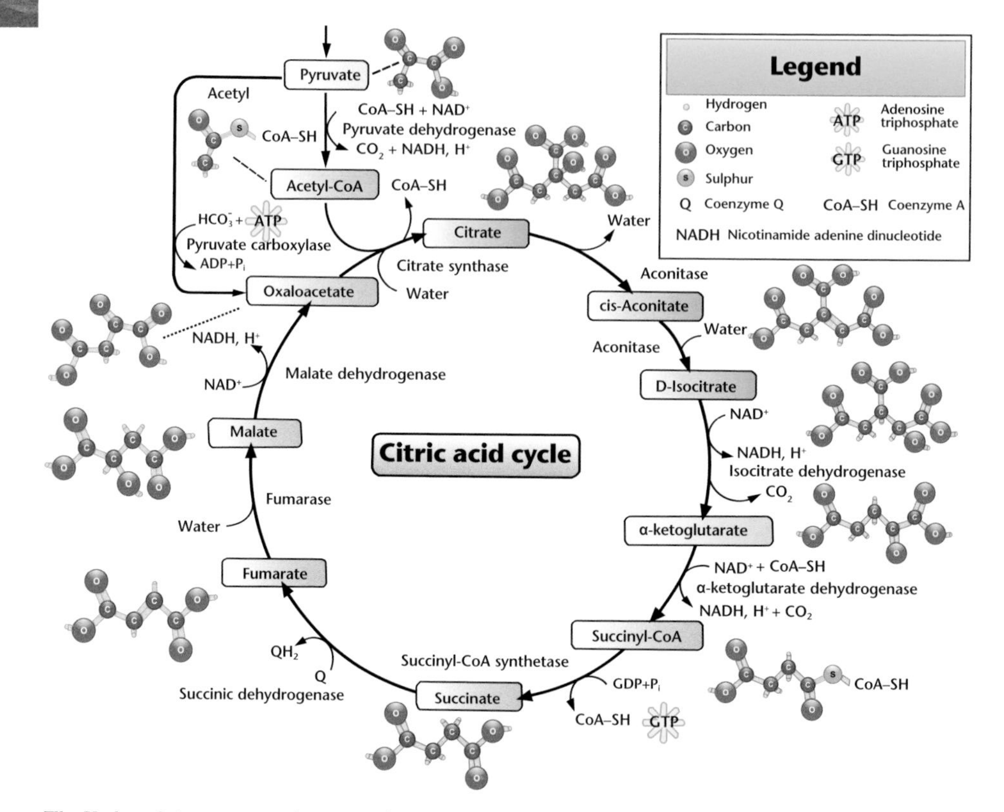

The Krebs cycle is a very complex series of reactions that takes place in all cells. It is part of an even more complex procedure that enables organisms to harvest energy from food.

are used? Where do they come from? Where do they end up? How does the multitude of chemical reactions in a body perform life processes?

The first person to discover a full chemical cycle was the German-born chemist Hans Krebs. Working in England in 1937, he unpicked the citric acid cycle – sometimes called the Krebs cycle in his honour. This describes the metabolic pathway by which organisms derive and store energy. It starts with acetyl-CoA and, through a series of reactions driven by enzymes, energy is harvested from the bonds in its molecule and stored for use in the cell.

More important than the details of the Krebs cycle for our purposes is its status as the first proof that all life processes can be reduced to a sequence of chemical reactions. Since 1937, many more metabolic pathways have been traced. They divide into those that release energy by breaking down complex molecules (catabolic pathways) and those that use up energy to build complex molecules (anabolic pathways). The details of the biochemistry are beyond the scope of this book.

A long path

Although the Krebs cycle was the first cycle to be unravelled, the first metabolic

Spallanzani (see page 154) persuaded (probably reluctant) birds to swallow capsules attached to a thread so that he could retrieve them to investigate the process of digestion.

pathway to be explored was glycolysis. This is the pathway that turns glucose (a sugar) into pyruvate; pyruvate is the starting point for the Krebs cycle. It took 100 years to piece together this pathway – partly because no one was looking for it at the start. The journey began with Pasteur's discovery that fermentation is performed by yeast and was put on track by Buchner's discovery that the living yeast organism is not needed.

From 1905–11, Arthur Harden and William Young measured the rise and then fall in carbon dioxide levels when yeast and glucose were warmed together. They found that adding an inorganic phosphate restarted the process, leading them to work out that a product of the process was organic phosphate esters. With a bit more work, they extracted fructose 1,6-diphosphate. Others continued to piece together the complex pathway, discovering which enzymes were responsible at each stage for catalyzing reactions until, in the 1930s, the German physiological chemist Gustav Embden outlined the main stages of the pathway. The final pieces were put into place in the 1940s.

Over the course of the 20th century it became increasingly clear that living

organisms are really communities of proteins working together at an incredibly complex series of tasks and pathways, from haemoglobin carrying oxygen in the blood to messenger RNA interpreting and implementing genetic instructions and neurotransmitters carrying nerve signals to and from (and within) the brain. The activities are purely chemical and can, theoretically, all be duplicated in the lab given sufficient expertise and resources. We're not there yet, but there is no vitalist magic or spirit needed to make the chemistry of life work.

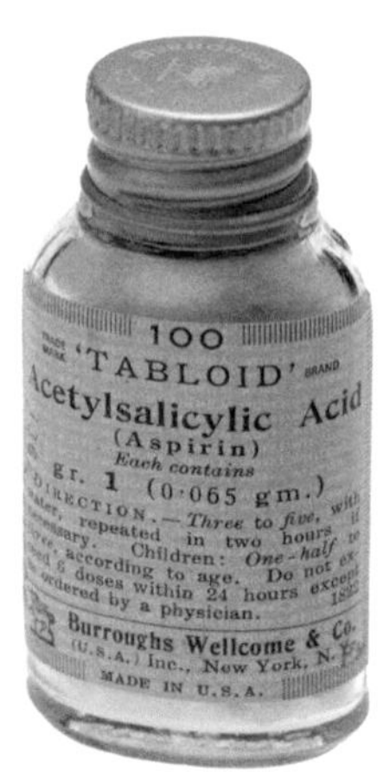

A bottle of aspirin produced by Burroughs Wellcome & Co, New York, in the 19th century.

Cures from chemistry

Just as chemistry is at the heart of running a living functioning body, so it also offers a way to treat one that is functioning poorly. The use of inorganic chemicals to treat illness can be dated back to Paracelsus and the other early iatrochemists, but most medicines have been derived from plant or animal sources. At the end of the 19th century, for the first time, a medicine previously extracted from a plant was synthesized in the laboratory. That medicine was aspirin – acetylsalicylic acid, a modified form of the salicylic acid found in willow bark.

Willow bark, Napoleon, and the first mailshot

Medications made from willow and other plants rich in salicylates had been used since Sumerian times, 4,000 years ago. They were used to treat fever and pain in Ancient Egyptian, Greek, Chinese, Roman and medieval European medical practice. A study of willow-bark extract in 1763 by Reverend Edward Stone showed what everybody already knew: that it was effective against aches, pains and fevers.

Further chemical investigation began the following century, prompted at least in part by Napoleon Bonaparte. Quinine was the preferred fever medication, and was brought from Peru, but in 1806 Napoleon's naval blockades had cut off the supply. European chemists set about finding an alternative, and explored variants of salicylic acid. In 1828, Johann Büchner,

THE HEROIC WONDER DRUG

Bayer was not initially keen to expend a lot of effort on developing salicin, since the stomach problems it caused seemed a serious drawback. Instead, they turned their attention to an alternative painkiller they were developing. The new drug was diacetylmorphine, and trial subjects reported that it was great, it made them feel 'heroic' – so in 1874 Bayer named it 'heroin' and advertised it as a safe, non-addictive alternative to morphine. It was sold principally as a cough suppressant, especially for children. A few years later, problems began to emerge. It was not as 'non-addictive' as originally claimed, and some people resorted to selling their old junk for money to feed their habit. Those people were called 'junkies'.

a professor at the University of Munich, isolated salicin, as he called it (from the Latin for 'willow'), from willow bark. The following year, French pharmacist Henri Leroux isolated a pure crystalline form of salicin and used it to treat rheumatic pain. But there were problems: pure salicylic acid irritated the stomach, leading to problems including bleeding and vomiting. Felix Hoffman, working for the drug and dye firm Bayer, was given the task of finding an improvement. It was a task he took on readily as his father suffered severe rheumatic pain and could no longer tolerate the stomach problems caused by salicylic acid. In 1897, Hoffman found that if he modified salicylic acid slightly to make acetylsalicylic acid, it was much more readily tolerated. He didn't know the reason at the time, but it seems that acetylsalicylic acid is easily absorbed, then converted back to salicylic acid in the body. Or that's the usual story. In 1949, an ex-employee of Bayer, Arthur Eichengrün, claimed that he had invented aspirin and Hoffman had been operating under his instruction. For a long time, the claim was dismissed, but in 1999 researcher Walter Sneader upheld Eichengrün's version.

Illustration from 'Experiments and observations on the Cortex Salicis Latifoliae, or Broad-leafed willow bark', 1803, exploring the efficacy of willow bark in combatting fever.

BAD LUCK WITH BLOOD AND DRUGS

The manipulative and influential Russian mystic Grigory Rasputin was called in by the Russian Imperial family in 1905 to try to treat the sickly prince Alexis. Horrified to find the boy treated with a synthetic chemical remedy, aspirin, he stopped the treatment and instead applied more 'mystical' treatments. The mystical treatments probably did Alexis no good at all, but stopping the aspirin improved his condition immensely as he was haemophiliac (meaning his blood could not clot). Aspirin simply made his blood problem worse. Rasputin's success in treating the tsarovich contributed to the tsarina's confidence in and devotion to him, increasing Rasputin's power within the Russian imperial household.

Even though Hoffman/Eichengrün and colleagues had not been supported by Bayer, they continued to develop their new acetylsalicylic-acid-based drug, which was just as well since Bayer soon needed it when heroin ran into problems (see box on page 158). Bayer released it under the trademark 'Aspirin' in 1899, marketing it with the first mass-mailing in history (to 30,000 doctors). It was also, soon after release, the first drug to be sold in tablet form. It was seen as a wonder-drug – an easy, cheap treatment for aches and pains and fevers, with none of the addiction problems of heroin. It was soon popular (and manufactured) worldwide, its position unchallenged until paracetamol (acetaminophen in the US) was introduced in 1956. Indeed, aspirin became even more of a wonder-drug in 1953 when a GP in California noticed that none of his patients taking it had suffered heart attacks.

That was explained in 1971 when the English pharmacologist John Vane discovered how aspirin works. It suppresses production of prostaglandins, which are involved both in sensitizing spinal neurons to pain and in the contraction and dilation of muscle. But it also suppresses production of thromboxanes, which – along with prostaglandins – are involved in blood clotting. Blood clots forming in the arteries of the heart are a common cause of heart attacks.

Paul Ehrlich's concept of the 'magic bullet' was a drug designed to affect only a specific, targeted disease vector.

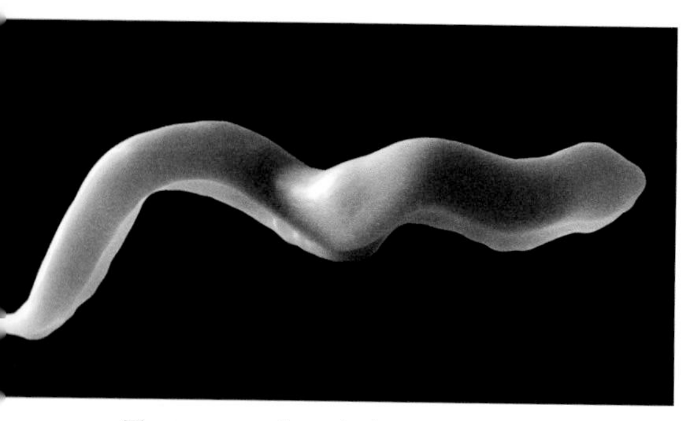

Trypanosoma brucei, *the parasite that causes sleeping sickness, a deadly disease endemic in sub-Saharan Africa (see opposite).*

Magic bullets

Aspirin was made by synthesizing a substance that occurs naturally and had been used for millennia, though changing its composition slightly. The next step was to produce

DON'T GO THERE

In 1939 Domagk was awarded the Nobel Prize in Medicine for his discovery of Prontosil but was forced by the ruling Nazi party to refuse it, and was imprisoned for a week by the Gestapo. German nationals were forbidden to accept Nobel prizes after Carl von Ossietzky, a critic of the Nazis, had been awarded the Peace Prize in 1935. Domagk finally received his Nobel medal in 1947, but lost the cash component of the prize as too much time had elapsed between the award and its acceptance.

medicines which had little connection with naturally occurring chemicals.

During the second half of the 19th century, chemists made and tried out lots of organic compounds, often discovering that they killed the agents that caused disease but did so much damage to the human body that they were not useful as medicines. One such substance was atoxyl (arsanilic acid). This was found to kill the *trypanosome* parasite which causes sleeping sickness, but resulted in blindness for the patient.

The German chemist Paul Ehrlich had an inspired idea. He was interested in exploring the uses of chemicals in disease control, and it occurred to him that the part of a molecule which does the good work (acting to kill parasites, for example) might not be the same as the part that does the harm (such as making people blind). In 1906, he came up with a plan: to make lots of variants of the molecule and try them all out – perhaps one would cure sleeping sickness without leaving the patients blind. He won the Nobel Prize in Chemistry for this concept, which he called the 'magic bullet', in 1908. Ehrlich worked out the molecular structure of atoxyl and began making modified versions which he tested on mice. By 1909, he had found a

A kit for administering Salvarsan by injection, 1912.

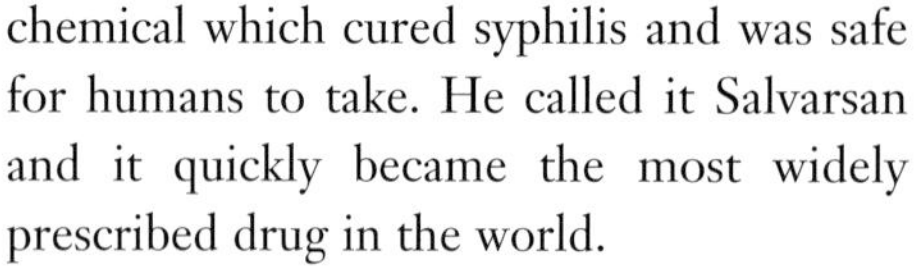

chemical which cured syphilis and was safe for humans to take. He called it Salvarsan and it quickly became the most widely prescribed drug in the world.

The German bacteriologist Gerhard Domagk followed the same principle. In 1932, he was testing sulphonamides on laboratory mice infected with bacterial infections. He found that one, Prontosil, was particularly effective. Before he had the chance to test it on humans, his own daughter fell seriously ill. No drug made her better. Taking a chance, he gave her a dose of Prontosil and, to his delight, she was cured. He then organized clinical trials and the drug was approved and immediately successful. But Prontosil is a complex molecule and difficult to manufacture. In 1936, work at the Pasteur Institute in Paris found that the effective part of Prontosil is sulphanilamide, and this soon overtook Prontosil.

Salvarsan, Prontosil and sulphanilamides saved many lives, but 'magic bullet' medications were overtaken in the 1940s by antibiotics, another accidental discovery, and one that returned the role of manufacture to natural organisms.

Fleming's fungus

The story of the discovery of the antibiotic drug penicillin is well known. The Scots biologist and pharmacologist Alexander Fleming had been growing cultures of the bacterium Staphylococcus on agar plates when he went away on holiday, piling them up for disposal. On his return, he found a

FROM ACETONE TO ISRAEL

Penicillin was not quite the first example of chemical manufacture being accomplished by a biological process on an industrial scale. That honour goes to the production of acetone ($(CH_3)_2CO$) during the First World War. Acetone is essential to the manufacture of cordite, an explosive more powerful than gunpowder. The Russian-born chemist Chaim Weizmann developed a process for producing acetone from the fermentation of glucose or starch by an acid-resistant bacterium, *Clostridium acetobutylicum*. The Weizmann process was soon scaled up to produce acetone for the allied forces and made a significant contribution to their military success.

In 1917, German submarine activity curtailed the supply of maize to Britain from the USA (one of the sources of starch used), and British-grown grain and potatoes were needed as food. School children and boy scouts were urged to collect acorns and conkers – the fruit of the oak and horse chestnut tree respectively – for processing into acetone. Around 3,000 tonnes of conkers were collected and moved by train to secret factories.

Weizmann was also an enthusiastic and leading Zionist who spent considerable effort influencing political figures in Britain (his adoptive homeland) in favour of the formation of a Jewish state. The importance of his acetone-producing process helped him gain sympathy for his cause. When the state of Israel was founded, Weizmann's campaigning and contribution to the war effort were recognized. He became the first president of Israel at the inauguration of the state in 1949.

Penicillium mould growing on an agar plate.

policeman, back from the brink of death only to lose him a few days later when they ran out of penicillin. But its effectiveness had been proved, and soon penicillin was in production and saving the lives of casualties in the Second World War.

clear space on one plate where bacteria had been killed by a mould that had grown on it, *Penicillium notatum*. Further investigation showed the 'mould juice' (as he referred to it) could kill other types of bacteria, including streptococcus, meningococcus and the diphtheria bacillus. He set his assistants the task of isolating the active substance that the mould produced. It proved difficult to extract and was chemically unstable, so although Fleming published his results in 1929 it was ten years before penicillin became a usable drug.

The work to turn mould juice into medicine was carried out in Oxford by Howard Florey and Ernst Chain, originally from Australia and Germany respectively. They developed their laboratory into a factory producing penicillin, growing the fungus broth in bath-tubs, milk churns and bed-pans. Their first trial, in 1941, was both encouraging and disappointing: they brought their first patient, an injured

When the body's chemistry goes wrong

Some forms of illness are caused by invading microorganisms, such as the bacteria treated by penicillin, but others are the result of the body's own chemical systems malfunctioning. As discovered in the 1950s, all body processes revolve around the manufacture and operation of proteins. The number of proteins in the human body is thought to be around 20,000 at any one time from a total repertoire of up to a million, so there's quite a lot of potential for something to go wrong.

Sweet and deadly

One of the enzymes discovered in the 19th century was insulin, which is produced by the pancreas and is important in regulating the absorption of sugar. In 1889, the Polish-German physician Oskar Minkowski removed the pancreas from a dog to test its function. Soon flies were seen swarming around the dog's urine, which was sweet. It

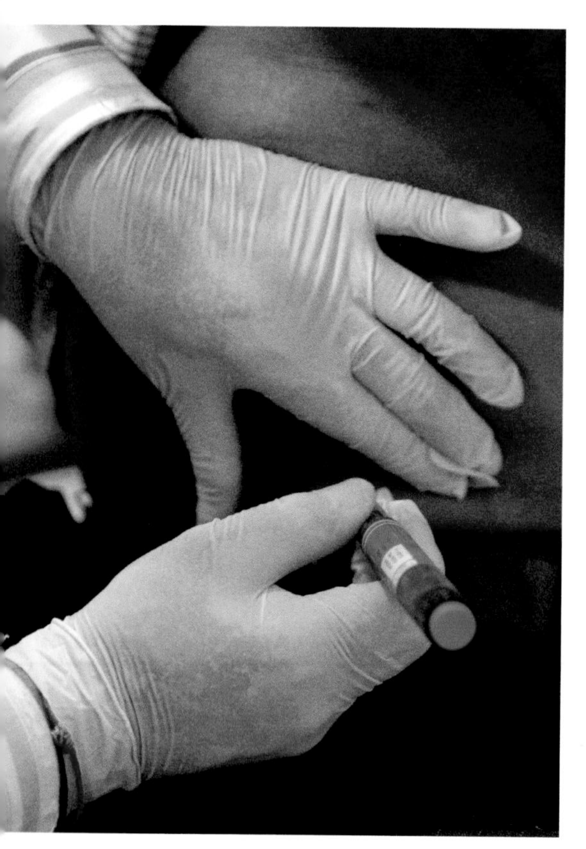

Insulin injections help people with diabetes to manage their condition.

became clear that the pancreas is involved in the absorption or excretion of sugar. Even so, it took until 1921 to extract insulin from the pancreas and prove its role, leading to the discovery that diabetes can be managed by administering insulin.

At first, insulin could only be harvested from the pancreas of other animals, usually pigs. But the chemical structure of two forms of insulin was discovered in 1951 and 1952 by British biochemist Frederick Sanger; the first synthetic insulin was produced a decade later. Today, most insulin used to treat diabetes is bioengineered: it is produced by yeast or a bacterium (*Escherichia coli*) that has been genetically engineered to produce human insulin. The technique was first used in 1978. Manufacture of bioactive molecules has been handed back to living organisms; but this time we're in charge, and the organisms have been redesigned as chemical factories.

Chemicals from the past

Of the two natural sources of organic compounds, living organisms and fossilized deposits, the second is most easily exploited in volume. Coal, oil and natural gas are all products of the fossilization or decay of organic material. Coal and oil are produced over millions of years: the coal we burn now started out as living trees 300 million years ago. Natural gas – mostly methane – is produced by the decay of organic material buried deep in the ground or by the action of microorganisms in bogs, swamps and shallow sediment.

Perkin discovered his mauve dye while experimenting with coal tar in his attempt to make quinine. Many more useful substances would come from working with tar and oil in its various forms, a number of them also discovered accidentally (see page 196).

Underground treasure

'Black gold' is not a new discovery. Various types of fossil fuel have been known for millennia in some parts of the world, and their chemical processing began long ago. Bitumen is a thick, black, sticky form of petroleum. It was discovered and used for making walls and towers in Babylon 4,000 years ago. The first oil wells were drilled

Bubbles of methane gas released by plants are trapped under the ice of Abraham Lake, Canada.

in China before AD347 and, by the 10th century oil was being carried by bamboo pipelines. The oil was used as a fuel; it was burned to evaporate water from brine to produce salt. In 7th-century Japan, petrol was known as 'burning water'.

Fossil fuels are a complex mixture of chemicals including compounds that are huge hydrocarbon chains, incorporating branches and benzene rings. They can be separated approximately by distillation.

Fraction by fraction

Although the alchemists used distillation extensively, the process they employed involved distilling the same substance repeatedly to purify it. The earliest distillation known is the simple process of boiling a liquid and collecting a single condensate. A basic still dating from 3600BC found in Mesopotamia comprises a large pot of about 40 litres capacity and a collecting ring that could hold two litres. It might have been used to make perfume. The Arab chemist Ibn Sina made the first modern perfume using steam distillation in the 10th century, mixing water and rose petals, heating them together and collecting the steam distilled as rose water.

Fractional distillation, which separates substances with different boiling points, was developed in the late 18th century. It

FRACTIONAL DISTILLATION

Fractional distillation is carried out in a column that is hottest at the bottom and coolest at the top. The mixture is heated at the bottom and vaporized. The gases rise up the column, those with the highest boiling point condensing out first (low in the column) and those with lower boiling points continuing to rise as vapour until condensing at lower temperatures further up the column. The process does not neatly separate compounds one by one. Each fraction can contain several different compounds with a similar boiling point.

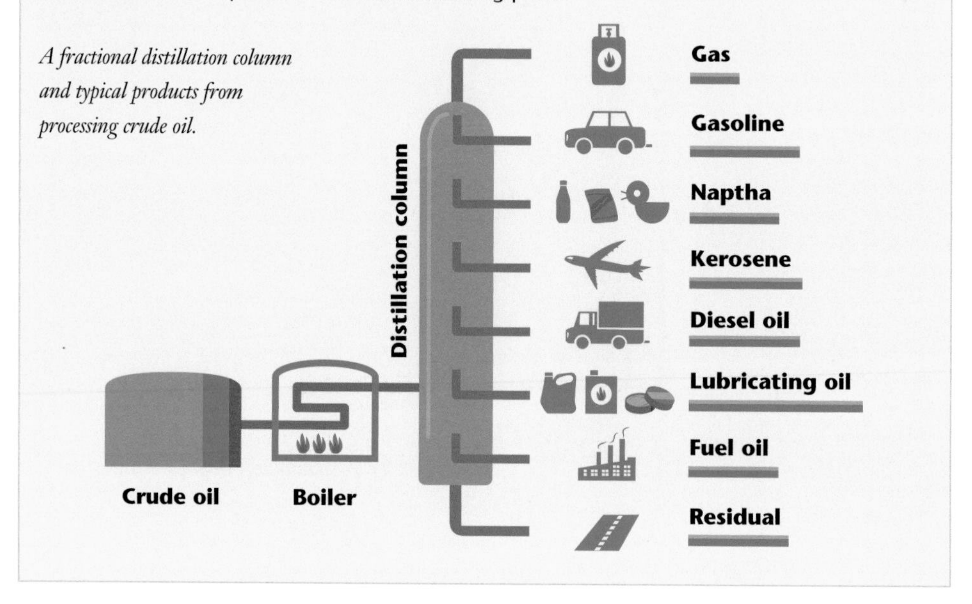

A fractional distillation column and typical products from processing crude oil.

relied on physical laws discovered by Dalton and Raoult relating to the partial pressures of gases in a mixture. Dalton's law states that the total pressure exerted by a mix of gases is equal to the partial pressures of each component. Raoult's law says that the fraction of the total pressure produced by each component is relative to the molar proportion of the component.

A new industry

For centuries, people have discovered oil and even petroleum seeping from rocks or bubbling up from the ground. Perhaps the start of the modern petrochemical industries should be dated to 1710 or 1711 when the Russian-born Swiss physician Eirini d'Eyrinys discovered asphaltum and developed a bitumen mine near Neuchâtel, Switzerland. His mine operated until 1986.

The first oil well and refinery was built in Russia in 1745 by Fiodor Priadunov, who produced kerosene by distilling petroleum, using it to fuel monastery lamps. But this was a small scale operation. A breakthrough came in 1847 when Scots chemist James Young found natural petroleum seepage in Derbyshire and discovered that he could

ACCIDENTAL ROADWAYS

The use of tarmacadam for road surfaces dates from a minor industrial accident in England in 1901. The civil engineer Edgar Hooley was passing a tar works when he noticed that someone had spilt a barrel of tar on the road; to cover the mess, gravel had been dumped on top of it. Hooley noticed that the part of the road covered with tar and gravel was dust-free. The following year, he developed and patented tarmacadam as a product for road surfacing.

Workers laying tarmacadam in London, 1920s.

distill it to produce a thin lamp oil and a thicker oil useful for lubrication. In 1850 he patented processes for deriving several substances from coal, including paraffin (a more refined form of kerosene). He went into a partnership and opened the first commercial oil works in the world, in Scotland. Meanwhile, a Canadian geologist, Abraham Gesner, discovered how to derive kerosene from coal, shale oil and bitumen. This led to kerosene lighting in cities, first in Canada and then in New York and other American cities. The Polish pharmacist Ignacy Łukasiewicz refined the method and managed to extract kerosene from petroleum, which was much more readily available.

It was, then, the distillation of kerosene, used first in street lighting, that set us on the path to using liquid fossil fuels and natural gas. The invention of the internal combustion engine and development of the motor car soon provided further demand for petroleum products. But there was more than fuel to come from petroleum and oil. Fossil fuels also yielded, again by accident, products that became central to 20th century consumerism – plastics and artificial fibres, as we will see in chapter 9.

CHAPTER 8

WHAT'S IN THERE?

'There is no more certain genus of acquiring knowledge than when one knows what is contained in a thing and how much of it there is.'

Jan Baptist van Helmont, 1644

The most important step in investigating a mixture or a compound is to find out what exactly is in it. The task of investigating the composition of a chemical is called analysis. Analytical chemistry has been important since the earliest days of chemistry. Consequently, some of the techniques used are very old. But the work is not finished when the chemist has a list of elements. As we have seen, how the atoms in a compound are arranged is also important.

Chemical analyst working in quality assurance in a brewery in Mongolia.

Investigation and identification

Today, analysis is carried out for many reasons: to ensure the quality or purity of a substance, to reveal its composition so that it can be replicated and to investigate problems, including contamination and pollution. The last includes forensic science, such as identifying poisons, detecting accelerants used in cases of suspected arson and investigating the fraudulent copying of trademarked products such as cosmetics and foodstuffs.

Chemistry, wet and dry

Traditional methods of analysis are often referred to as 'wet' or 'dry'. Wet methods include chemical tests – for example, adding chemicals to a compound that change colour in the presence of a particular substance – and are the typical fare of the laboratory bench, familiar to many people from school science lessons. Most modern analytical laboratories have replaced or augmented these methods with sophisticated instruments. Since the mid-20th century, this work has become

GREEN AND DEADLY

Arsenical compounds make effective green dyes, but they are also highly toxic. In the past they were used to colour paint, wallpaper and even foodstuffs. In the 19th century, a green sugar cake decoration made with arsenic and sold in Greenock, Scotland, poisoned several children.

In 1858, in Bradford, England, the accidental use of arsenic in a batch of humbug sweets caused 21 deaths and over 200 further cases of severe illness. The arsenic compound had been mistakenly sold as a sugar substitute rather than as a dye. That catastrophe led to the passing of the UK's Pharmacy Act in 1868 to help control the sale of poisonous chemicals.

The dangers of using arsenic in confectionery are lampooned in a 19th-century cartoon. The box of Plaster of Paris alludes to another, less dangerous form of adulteration.

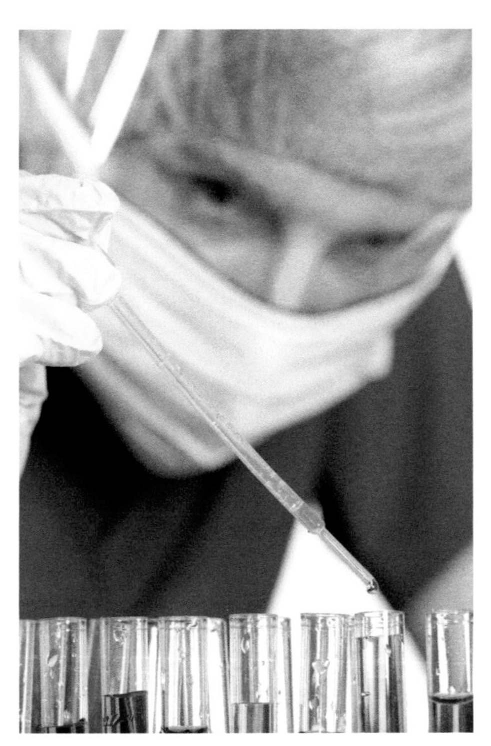

Analysis still involves detailed work by hand, often using equipment that has hardly changed in 1,000 years.

increasingly automated. 'Dry' methods originally used heat from a flame and are among the oldest.

Analysis and assay

Perhaps the earliest use of analysis was for assay, to assess the purity of metals. This is the oldest form of quality assurance, for example, to test the quality of gold.

Gold has been heated in a furnace both to test it and to purify it since at least 2600BC in Babylon. Any other metals mixed in with the gold would oxidize and fuse to the side of the crucible (see below), leaving pure gold. By weighing a sample of gold precisely before and after this treatment, it was possible to determine its level of purity.

The first standardized method of analysis dates from 1343 when Philip VI of France set out strict, detailed instructions on how the purity of gold should be tested. The method used was called 'cupellation', already in wide use before King Philip's involvement. The metal to be tested is melted with lead in shallow dishes called cupells. The lead oxidizes, drawing oxygen from oxides (that are impurities) in the metal. Metals with a lower melting point than silver or gold will solidify first as the mixture cools. The lead oxide is absorbed by the cupell itself. Philip VI stipulated that the cupells should be made of vine-shoot ash and the burnt leg-bone of a sheep. They must be shallow, well washed and polished, and then treated with liquid containing

A cupellation furnace, 1556. The method of cupellation was perfected in the 16th century.

ON METALS

In *De re metallica* (1556), Georg Bauer (also known as Georgius Agricola) devoted an entire book to processes for assaying and smelting ores and separating and testing metals. Generally, the process for assay was a small-scale version of the smelting process, measuring the ore and product. His book was the authoritative text on mining, smelting and the chemistry associated with metal-working for around 180 years.

Mining for metal ores, 1556.

suspended powdered deer-antler, which forms a white glaze and makes it easier to remove the material afterwards.

A new method of purifying or assaying gold was developed in the 15th century. A sample of gold tainted with silver and copper was melted with antimony, causing the impurities to react with the antimony and float on top of the gold.

Dry analysis of wet stuff

Another substance commonly requiring analysis was mineral waters. This was carried out using a dry method: heat from a flame was used to evaporate the water and leave the mineral solute behind. The chemist would examine the shape and colour of the crystals, taste and smell them, and perhaps blow them into a flame (sodium chloride was noted for crackling in fire) or heat them on a hot iron to observe the outcome.

The development of blowpipes in the second half of the 17th century enabled chemists to blow air into the fire, both to feed more oxygen into the flame and thus raise the temperature and to direct the flame more precisely towards the sample, which was often held on a charcoal block. After 1800, when English chemist William Wollaston perfected malleable platinum, the sample was sometimes held on a platinum wire for this process. Samples were frequently mixed with sodium carbonate or borax (a compound of boron, sodium and oxygen). With sodium carbonate, it might produce recognizable decomposition products; with borax, it might impart a characteristic colour to borax fused with glass, helping the chemist identify the sample.

EAT ME, DRINK ME

Although today's laboratory safety inspectorate would throw up their hands in horror at the prospect, smelling, feeling and tasting substances to help identify them was a common practice in analytical chemistry until the second half of the 20th century. This was even practised in medicine long ago, with physicians tasting a patient's urine to identify certain conditions. The name *diabetes mellitus* (from the Latin for honey, *mel*) comes from the sweet taste of the sufferer's urine, identified by Thomas Willis who named the condition in 1674. Tasting and smelling urine remained a common diagnostic tool until chemical tests became available.

A physician examining a urine sample, late 17th/early 18th century.

Wet gold

Modern analytical chemistry makes extensive use of wet methods; the first to appear was the use of mineral acids to dissolve and separate metals. As early as the 12th century, Albertus Magnus described the preparation of nitric acid, which was originally named 'separating water' because it could be used to dissolve the silver in an alloy of silver and gold. In the 15th century it became the principal method of separating these metals. Chemists soon discovered that it worked best if the ratio of gold to silver was 3:1, so they sometimes added silver to the mix in order to remove it all more efficiently. The mix of acids known as *aqua regia* (royal water), first made by Jabir ibn Hayyan in the 8th century, was the only liquid known that could dissolve gold. As it causes silver to precipitate out, it offered another wet method of separating silver and gold.

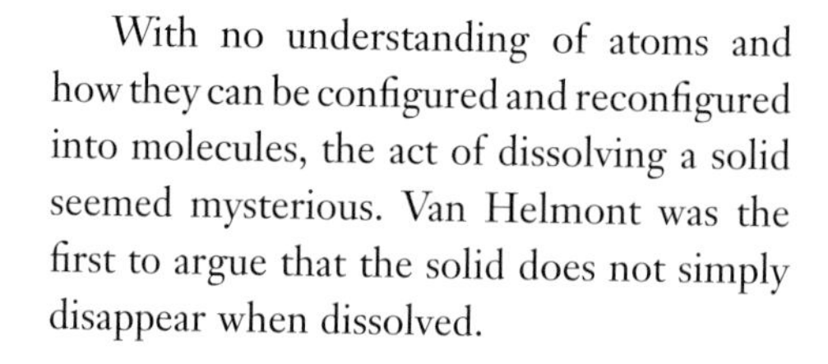

With no understanding of atoms and how they can be configured and reconfigured into molecules, the act of dissolving a solid seemed mysterious. Van Helmont was the first to argue that the solid does not simply disappear when dissolved.

Seeking the alkahest

Alchemists sought a substance they called the 'alkahest', a name probably coined by Paracelsus, that was capable of dissolving anything. There is an immediate problem, of course. If a liquid can dissolve anything, it can't be put into any type of container as that also will dissolve.

A more practical quest was to find an alkahest capable of dissolving any non-elemental matter. This was promoted by Eirenaeus Philalethes (or George Starkey, see page 73). Within the paradigm of the classical elements, a vessel made of earth could be used to dissolve non-elemental matter. Working within the modern paradigm, a container made of an elemental metal could be used to hold the reaction. Or at least, this would work if the alkahest actually existed. Paracelsus had a recipe based around caustic lime, alcohol and carbonate of potash. Van Helmont seems to have believed Paracelsus's recipe worked, calling it 'incorruptible dissolving water'.

Wet methods for wet stuff

The earliest known wet methods of analysis were used with mineral waters. The Roman writer Pliny noted in the 1st century AD that if a mineral water containing iron is dripped on to papyrus soaked in oak gall,

MODERN PH SCALE

The acidity or alkalinity of a substance is now measured using a pH scale of 1 to 14 with the midpoint, 7, representing neutral. Anything under 7 is acidic (with 1 being the most acidic) and anything over 7 is alkaline, or 'basic' (with 14 being the most alkaline). The simplest test is to use litmus paper, developed by the Spanish chemist Arnaldus de Villa Nova in the 16th century. This usually comprises strips of filter paper impregnated with dyes extracted from lichens. Litmus paper turns red in an acidic environment, blue in an alkaline one and purple when the conditions are neutral. The shade shows the relative acidity/alkalinity and can be compared with a reference colour chart to give the pH value.

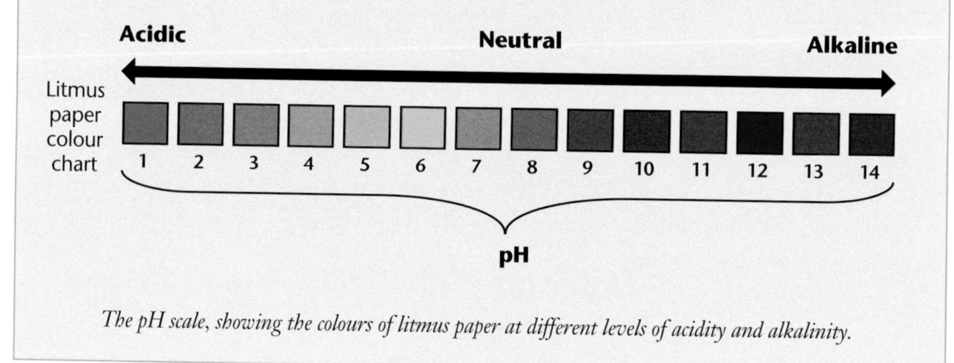

The pH scale, showing the colours of litmus paper at different levels of acidity and alkalinity.

the drops turn black. This was apparently used to determine whether copper sulphate was adulterated with iron sulphate. The method is not mentioned again in Europe until Paracelsus used it in 1520.

Robert Boyle was first to note how some vegetable juices could be used to indicate the pH (acidity or alkalinity) of a substance. He recorded in his *Experiments and Considerations Touching Colours* (1664) that syrup of violets and extract from privet berries (and many other blue extracts from plants) will turn red in the presence of acid, and green in the presence of alkali, but remain unchanged in the presence of neutral substances. The modern pH scale was still a long way in the future (see box, opposite).

Boyle, however, was satisfied with discovering that – contrary to popular belief – substances need not be either acid or alkali but can be neutral. The concept of pH was introduced by the Danish chemist Søren Sørensen at the Carlsberg brewery laboratory in 1904 and revised to the modern scale in 1924.

Testing, testing

There are many tests to reveal the presence of different chemicals. They have been developed over time as the need arose. At first, they were discovered accidentally, like the Roman test for iron using oak gall. Later, chemists used their knowledge of 'affinities' to work out tests which would exploit the reactivity of elements, using one to replace or change another.

An example of necessity driving chemistry is the Marsh test for arsenic. Arsenic is highly toxic, but in the 19th century it was widely used and readily available to kill pests and vermin. Inevitably, it was also a favourite of poisoners. It takes little arsenic to kill someone, as shown by the Bradford poisonings (see page 170). James Marsh was a Scots chemist called upon to give evidence in the trial of a suspected poisoner in 1832. John Bodle had been accused of killing his grandfather by giving him coffee laced with arsenic. Marsh used the standard test for arsenic and found it present, but by the time of the trial the product that proved the case had deteriorated and the murderer was acquitted. This miscarriage of justice spurred Marsh to devise a better test for arsenic – one that can detect as little as a fiftieth of a milligram. He combined the sample with sulphuric acid and zinc, producing the gas arsine (AsH_3). When lit, the hydrogen burns and arsenic is deposited as a solid on a cold surface.

A demonstration of the chemistry of arsenic, 1841.

Unmixing stuff

To find out what is in a mixed substance, the components are usually separated. Methods of separation include filtration, precipitation, distillation, extraction and chromatography. All except the last of these are very old methods.

Filtration is used to separate solid particles from a liquid, for example, by pouring water through filter paper or cloth; the solid particles are left behind. This method has been used since prehistory. Precipitation is used to separate out a substance that is in solution. Chemicals may be added that act on the dissolved substance, which then becomes a precipitate and might stay suspended in the liquid as particles, or float on the surface, or sink to the bottom of its container. Distillation is used to separate a liquid from a solid by heating, or for making a more concentrated solution, or for separating liquids with different boiling points (see page 165). As a liquid reaches its boiling point it becomes a gas and separates from the mixture. It is then cooled, and condenses into a liquid again.

Extraction works by dissolving part of a mixture. The mixture is added to a solvent capable of dissolving only some of the components of the mixture. A common example is in making tea: the tannins, theobromine, polyphenols, and caffeine in the tea leaves are soluble in hot water, but other components, such as the cellulose of the tea leaves, are not soluble. This method has been used since prehistoric times in cooking and in making dyes, medicines and perfumes. Solutes can be moved between solvents, too, as long as the solvents are immiscible (won't mix together). Usually, one solvent is water and the other an organic solvent. Some of the solute will move out of aqueous solution (the water) to dissolve in the organic solute.

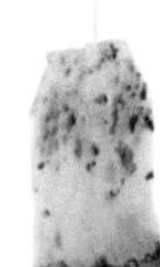

Using a tea-bag, only the water-soluble parts of tea pass into the cup.

Precipitation explored

Robert Boyle actively explored precipitation as a method of separation. At the time, it was believed that an 'antipathy' between substances caused one to precipitate out of a solution when another substance was added. For example, a metal thrown out of an acidic solution by the addition of alkali was considered to result from the antipathy between acid and alkali. Boyle demonstrated that sometimes a precipitate could be formed using a neutral precipitant, such as when common salt forces silver to precipitate from solution in nitric acid. Finding that if he dried the precipitated silver it weighed more than the silver he had originally dissolved, he concluded that the silver and the precipitant formed a 'coalition'.

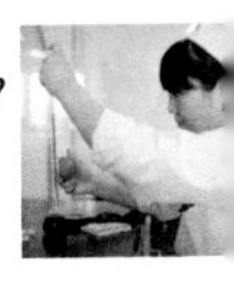

Over the coming centuries, more precipitants were discovered and wet analysis increasingly took over from dry methods. Although initially a qualitative approach, wet analysis became quantitative as chemists learned to perfect their extractions and weigh them, then calculate the percentage of the substance by weight in the original sample. This gravimetric (by mass) analysis was particularly useful in assessing ores. Several new metallic elements were discovered in this way. The German chemist Martin Klaproth used this method to identify uranium, zirconium and cerium.

When silver nitrate is added to a solution of potassium chromate, an orange solid (silver chromate) precipitates.

Analysis became more formalized and structured in the 19th century, with the first text on analytical chemistry published in 1821 by Christian Pfaff. He described each reagent, explaining how to prepare and use it. At the time, reagents were all inorganic, but by the end of the 19th century synthetic organic reagents were added to the analyst's toolbox.

QUALITATIVE AND QUANTITATIVE ANALYSIS

Qualitative analysis aims to find out the components of a substance. It does not attempt to find the proportions of each. Quantitative analysis is concerned with the quantities of each component in a substance.

Titration

Weighing the precipitate produced in a reaction is one way of judging the amount of a substance present in a sample; another way is to measure the amount of a reagent that is used to produce a reaction. This is called titration and was first mentioned as a method in 1729, though in the very first case Claude Geoffroy (1729–53) weighed the reagent instead of measuring its volume. He found a way of measuring the acid content of vinegars by adding small amounts of potash and observing the consequent

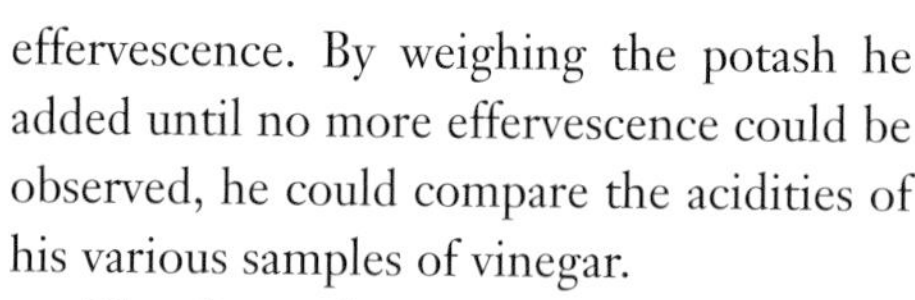

effervescence. By weighing the potash he added until no more effervescence could be observed, he could compare the acidities of his various samples of vinegar.

The first volumetric titration recorded was carried out by Francis Home in Scotland in 1756. Using the same reaction, but the other way round, Home added nitric acid to potash, one teaspoonful at a time, until the reaction was complete. It was imprecise, but it established the volumetric method of measurement.

Titration was greatly improved by the development of indicator solutions that change colour to show the endpoint of a reaction. William Lewis wrote in 1767 of using the change in colour of 'certain vegetable juices' to indicate the point of saturation by an alkali. In the 1850s, Karl Schwarz used sodium thiosulphate to indicate iodine liberation from potassium iodide by oxidizing agents. Potassium permanganate was first used as an indicator by Frédéric Margueritte in 1846 and synthetic indicators soon followed, beginning with phenolphthalein in 1877.

Spots and dots

Chromatography is the newest of the wet methods of separation and was first developed in Russia by the Russian-Italian botanist Mikhail Tsvet (1872–1920) in 1900. Even before that, though, chemists had noticed that if a drop of a mixed liquid is put on to a filter paper, some of the components will spread out further than

Chemistry was an increasingly important part of manufacturing and commerce; this illustration from 1857 shows a lesson on industrial chemistry for workers in Paris.

In column chromatography, the components separate into different bands according to how far they can travel through the packing material.

others. The result is a set of circles, often of different colours, indicating which liquids have travelled further through the paper. Friedlieb Runge published his observations of the phenomenon in 1850 and 1855, and sometimes the method was used to compare samples of dyes and control for quality.

Tsvet developed chromatography to separate out the plant pigments (chlorophylls and carotenoids) that he was working on. He used a column packed with powdered calcium carbonate. He then poured the solution of plant extracts down it and noticed how different coloured bands appeared along the column. The different components in the mixture travelled different distances through the column, just as the components of a liquid mixture spread out on a filter paper to different distances. Tsvet could then wash out each separated substance using a solvent, collecting each component in turn.

More methods of chromatography have been developed since, but all work on the same basis of two different phases (solid and liquid, or liquid and gas), one carrying the sample through the other. Modern methods include gas chromatography, used with volatile substances that are carried in a mobile gas phase, and gel chromatography, in which a solution is carried through a gel phase.

Elements and electricity

Analysing a compound involves identifying the individual elements it contains, so accurate analysis wasn't possible until Lavoisier had managed to pinpoint some of the elements. Progress in identification and analysis went hand-in-hand. One particularly fruitful method involved the use of electricity.

Electricity put to work

Until the 19th century, the only way to split compounds into their constituent elements was to use chemicals or heat. But in 1800, news of the new 'voltaic pile' reached England. This was the first electric battery, devised in Italy by Alessandro Volta (see page 124). In 1791, he found that if he separated silver and zinc discs by cloth soaked in brine, and connected the metals with a wire, it caused an electric current to flow between them. This was the first electric cell. In 1800, he found that by stacking several such cells together he created a much more powerful device – the first electric battery. It was later found that copper worked just as well as silver and was far less expensive.

Within days of hearing about the new battery, William Nicholson and Anthony Carlisle had made one and used it to decompose water into hydrogen and oxygen, noticing that the two gases formed at different poles. By the end of the 19th century, Humphry Davy had worked out that a chemical reaction was taking place and that one of the metals was being oxidized. He found that the hydrogen appeared at the negative terminal (zinc) and oxygen at the positive terminal (silver/copper).

After further investigation, Davy concluded that chemical affinity is electrical in nature. On that basis, he theorized that it must be possible to break down compounds using electricity, because 'however strong the

Volta demonstrates his Voltaic 'pile' to Napoleon Bonaparte in 1801.

> **OVER AND OVER**
>
> Theodore von Grotthuss suggested that when hydrogen or oxygen was liberated at one of the terminals, showing that a water particle had been decomposed, the free gas would then recombine with the appropriate part of an adjacent water particle, displacing its equivalent – so a free hydrogen particle would displace a hydrogen particle from a nearby water particle. This set up a chain of successive decompositions and recombinations throughout the solution. This curious theory, which paid no heed to why particles might behave in such an energy-wasting fashion, persisted until the 1880s.

natural electrical energies of the elements of the bodies may be, there is every possibility of a limit to their strength; whereas the powers of our artificial instruments seem capable of indefinite increase'.

Consequently, he set about trying to use electricity to break down substances that had been resistant to previous attempts at decomposition, including potash and soda. By this means he eventually isolated potassium and sodium, both too reactive to remain as elements in nature. He went on to extract calcium, barium, strontium and magnesium in this way. Recognizing that the extreme reactivity of potassium might mean it could be used to decompose other substances, by displacing them in their combination with oxygen, he isolated boron from boric acid in 1808 – just a few

Although most often associated with electricity, much of Faraday's research involved chemistry.

days after Gay-Lussac and Louis-Jacques Thénard had accomplished the same feat. Davy's attempts to break down the elements nitrogen, sulphur and phosphorus were, inevitably, unsuccessful. But he did identify Lavoisier's 'oxymuriatic acid' as the element chlorine. Although he failed to isolate fluorine, he was confident that he had found another element, this time in hydrofluoric acid. In 1813, he was given a sample of a purple solid obtained from seaweed and identified iodine. Gay-Lussac, whose researches paralleled those of Davy, had also found iodine and named it 'iode', a name that Davy simply anglicized.

Michael Faraday (1791–1867) first worked as an assistant to Humphry Davy but went on to work on his own investigations with electricity. He is most famous for discovering the underlying principles of electromagnetism, but also worked on the electrical decomposition of chemicals. He found that if he passed electricity through a solution of hydrogen chloride, the amount of hydrogen released depended only on the amount of electricity used. He also found that the quantity of different elements freed by the same amount of electricity is relative to their equivalent weights. This finding supported his theory that the electricity is interfering with the forces of 'affinity' that hold compounds together.

Sweetness and light

Although Lavoisier included 'light' in his list of elements, we now know that it is not an element. It has, though, played a role in identifying new elements and helping to determine the elements within a compound.

Breaking up light

When, in 1752, the Scots physicist Thomas Melvill experimented by throwing different materials into a flame, it was already known that some substances produce coloured flames. But Melvill found that if he passed the light from the flame through a glass prism, strange spectra appeared: there were dark gaps – sometimes quite large gaps – between the individual colours of the spectrum produced. The astronomer William Herschel developed this further in the 1820s, finding that he could identify samples of powdered elements by heating them in a flame and examining their spectra.

Mind the gaps

In 1802, the English chemist William Wollaston discovered that if sunlight was passed through a prism and the resulting

White light can be split into a spectrum of coloured light, as Isaac Newton discovered in the 1660s.

spectrum spread sufficiently widely, thin, dark bands appear. The German optical glassmaker Joseph von Fraunhofer made the same discovery two years later. Intrigued, von Fraunhofer investigated them systematically, finding an 'almost countless number'. They became known as Fraunhofer lines. Fraunhofer then began to study the spectra of stars and planets, comparing them with that of sunlight – a technique which has proved immensely valuable in astrophysics for discovering the chemical composition of the sun and stars.

Fraunhofer developed the diffraction grating, a series of closely-spaced slits that allowed him to measure the wavelengths of the spectral lines precisely, something not possible with a glass prism. Unfortunately, Fraunhofer died before he could take this work further. Around 30 years later, the German physicist Gustav Kirchhoff discovered that each element and compound has its own precise spectrum – a sort of fingerprint in the light it emits. Over the coming years, many people worked to determine the spectra of different light sources, learning to identify their chemical composition from their spectra.

CASTING LIGHT ON THE STARS

Spectroscopy enabled insights which had previously seemed impossible:

'Our knowledge concerning the gaseous envelopes [of stars] is necessarily limited to their existence, size . . . and refractive power, we shall not at all be able to determine their chemical composition or even their density. . . . I regard any notion concerning the true mean temperature of the various stars as forever denied to us.'

Auguste Comte, 1835

Light in and out

In 1848, the French physicist Léon Foucault discovered absorption spectra which, as the name suggests, are the opposite of emission spectra. He realized that if he placed a strong light behind a flame containing sodium, the yellow band in the spectrum was absorbed. Kirchhoff worked the two discoveries together in 1859, finding that a substance will absorb light of the same wavelength that it emits. If absorption and emission spectra are placed together, the gaps in one match the coloured bands in the other.

Kirchhoff worked with the German chemist Robert Bunsen (1811–99) to catalogue the spectra of thousands of substances to an accuracy of 0.01 per cent. Between 1855 and 1863, they logged spectra by introducing salts into the flame of a Bunsen burner (which Bunsen had just invented) to produce emission spectra, and used a cool flame of burning alcohol to study absorption spectra. In the course of their investigations they found two new elements, rubidium and caesium; other chemists discovered a further 15 new elements using this technique, now known as spectroscopy.

Kirchhoff and Bunsen explained that the Fraunhofer lines found in sunlight were caused by elements on the surface of the Sun absorbing light emitted by the hotter interior, and so enabled the chemical analysis of the Sun's atmosphere. The Swedish chemist Anders Ångström had observed the spectrum of hydrogen in 1853;

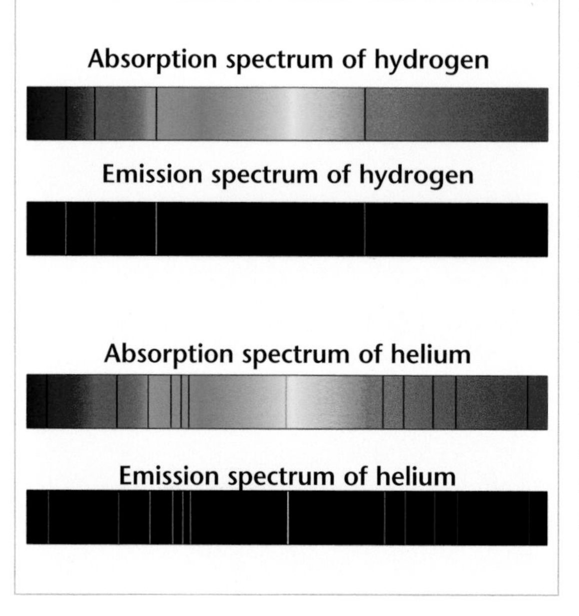

helium was discovered first in the Sun's spectral signature in 1868, the year before the element was found on Earth.

Kirchhoff and Bunsen established spectroscopy as an analytical tool, one that enabled chemists to identify the composition of a subject simply by burning it and matching the spectra produced to reference spectra listed in a catalogue. It is a qualitative rather than quantitative method – it reveals what is in a substance, but not how much of a component there is. Spectroscopy is the only way to determine the composition of stars.

While earlier analysts had to

BEYOND LIGHT

Spectroscopy can include infrared (IR) radiation emissions. An infrared spectrophotometer measures the wavelengths of infrared that are absorbed by a sample and these are then related to known bond energies. Different energy patterns are associated with different types of bond. The first IR spectra were published by the American physicist William Coblentz in 1905. He improved the technique and made painstaking measurements with equipment of his own design. He noted that some molecular groups produced characteristic spectra, which led eventually to the extensive use of IR spectroscopy for analysing organic compounds. It has a variety of applications in forensic science, from finding accelerants in cases of arson to detecting art fraud.

Ultraviolet
UVC UVB UVA
Visible
Infrared
100 280 315 400 700 wavelength (nm)

The electromagnetic spectrum, from ultraviolet to infrared.

ELEMENT ON THE SUN

Although we can produce light artificially, most of our light on Earth comes from the Sun. In 1868, the French astronomer Pierre-Jules-César Janssen was studying the spectrum of the Sun during a solar eclipse. He noticed an unfamiliar yellow spectral line. The English astronomer Norman Lockyer realized that the line could not have been made by any known element, so must represent an element present in the Sun but not on Earth. He named it 'helium', from the Greek for sun, *helios*. It was the first element to be discovered outside Earth.

Helium has since been found on Earth, where it is produced by the radioactive decay of rocks in the Earth's crust. The gas makes up only 0.0005 per cent of the atmosphere because it is constantly escaping into space. Though scarce on Earth, helium is the second most prevalent element in the universe. Helium is forged in stars by the nuclear fission of two hydrogen nuclei, the first stage of making all the other elements.

compare each spectrum obtained from a sample with known spectra, modern, computerized instruments handle this using a database of spectra. Techniques now extend beyond visible light to its close neighbours in the electromagnetic spectrum, such as infrared (see box, opposite), providing a wide repertoire of methods.

Measuring by mass

Imagine a ball is rolling along the floor and you direct a jet of water at it from a hose. The path of the ball will be changed, or deflected. How much it will be deflected depends on the mass of the ball and the power of the water jet. Using the same water jet, a lightweight ball will be deflected more than a heavy ball. The same principle can be applied to moving particles. This concept lies behind the mass spectrometer, another piece of equipment used to work out the composition of a chemical sample. It uses ions deflected in their passage through a vacuum; by measuring the deflection of the ion you can calculate its mass.

Investigating rays

Scientists started to experiment with discharge tubes in the mid-19th century. These are sealed containers in which an electric charge is passed through captive gas. The result is a stream of charged particles moving between the electrodes. In 1886, Eugen Goldstein found that with a perforated cathode (negative electrode), positive rays, which he called 'canal rays', move in a direction opposite to the cathode rays. J. J. Thomson identified cathode rays as a stream of

A discharge tube of the type used for cathode ray experiments.

electrons in 1897. In 1907, it emerged that the particles making up the canal rays were not all the same mass. This called for investigation.

In 1913, Thomson directed a stream of ionized neon in a discharge tube so that it travelled through a magnetic and an electric field. He measured its deflection using a photographic plate, which would produce a patch of light where the ions hit it. He found two patches of light, indicating that the particles were deflected on to two different paths. This could only mean that there were two types of particle with different mass – he had found two isotopes of neon (they turned out to be neon-20 and neon-22). His experiment formed the basis of mass spectroscopy.

Mass spectroscopy and isotopes

Thomson's student, Francis Aston, built the first mass spectrometer in Cambridge, England, in 1919 and used it to identify isotopes of chlorine, bromine and krypton. This finally explained the puzzling atomic weight of chlorine, 35.5 – it's an average of two isotopes. Aston went on to identify 212 of the 287 naturally occurring isotopes. This work led to the Whole Number Rule, which states that if the mass of oxygen is 16, all other isotopes have masses that are a whole number.

Aston also found that the mass of hydrogen is 1 per cent higher than expected (that is, than calculations based on the other elements would suggest). It represents energy that is lost when

X-ray spectrometer developed by Willam Henry Bragg and his son William Lawrence Bragg to investigate the structure of crystals, 1910–26.

hydrogen nuclei are forced together to make helium and then other elements (see page 135). In 1932, the equivalence of mass and energy implied by this finding and stated in Einstein's equation $E = mc^2$ was confirmed by Kenneth Bainbridge using his new, extremely accurate mass spectrometer.

Looking inside

Spectroscopy can identify the elements in a substance; mass spectrometry can identify elements, groups and isotopes; IR spectroscopy can identify the bonds in a compound (or mix of compounds). The next tool from the electromagnetic spectrum to be added to the analytical chemist's toolbox was X-rays. But instead of showing which substances are present in a sample, X-rays are used to investigate the structure of crystals, showing the positions of atoms. X-ray crystallography, invented in 1912, would give the vital clue that helped unravel the double helix of DNA and reveal the structure of other important biological proteins – the final part in the biochemical puzzle.

X-raying crystals

When the German physicist Wilhelm Röntgen discovered X-rays in 1895, others had just finished working out all the possible symmetries a crystal lattice could have.

SNOWFLAKES AND DIAMONDS

A snowflake is a composite of many snow crystals clumped together. Examined under the microscope, each snow crystal has hexagonal symmetry. X-ray crystallography shows the arrangement of water molecules in ice, revealing a tetrahedral arrangement of hydrogen bonds around each water molecule.

Another structure elucidated by X-ray crystallography is that of diamond. Diamond and graphite are both made entirely of carbon atoms, yet have very different properties, stemming from the different arrangements of the atoms in each structure. (Variant forms of an element with different arrangements of atoms are called allotropes.) The different arrangements of atoms in the allotropes of carbon are clearly visible using X-ray crystallography. In diamond, each carbon atom is covalently bonded to (that is, shares electrons with) four others in a giant matrix. In graphite, each carbon atom is only bonded to another three. The atoms are linked in a flat plane rather than a three-dimensional structure, with the result that graphite is composed of layers that easily slide over one another. You can write with a graphite pencil because it is very easy for a layer of graphite to slip off the pencil lead and on to paper.

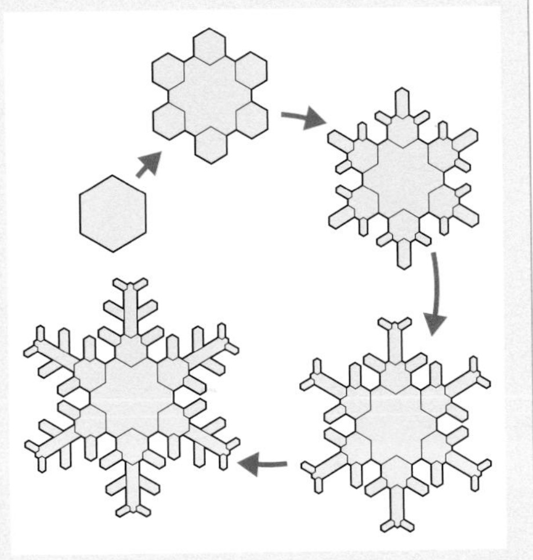

A snowflake builds symmetrically, following a pattern based on the hexagonal ice crystal at its heart.

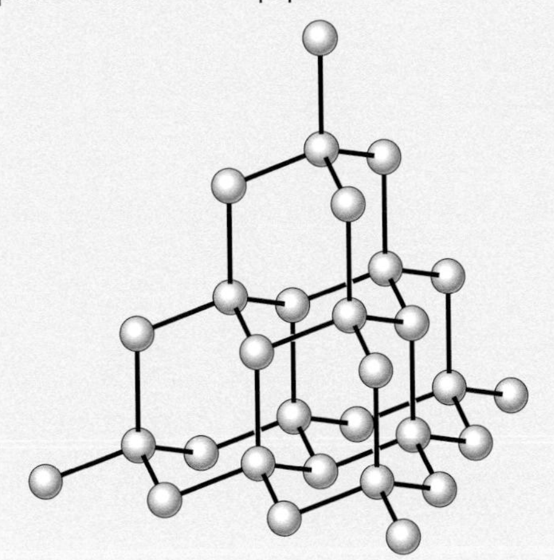

The crystalline structure of diamond, with equally strong bonds in all dimensions, gives the material its strength.

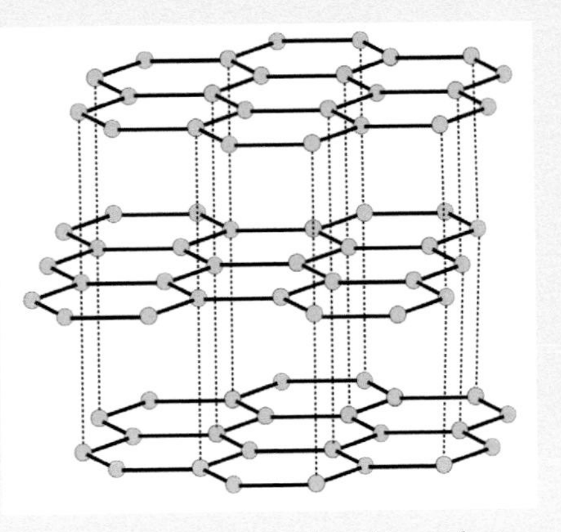

The crystalline structure of graphite, with atoms arranged in layers.

Putting the two discoveries together in 1912, Paul Ewald and Max von Laue shone a beam of X-rays through a copper sulphate crystal and recorded the diffraction pattern on a photographic plate. The result was a series of spots arranged in overlapping circles around a central circle representing the beam. With some genius mathematics, von Laue developed a law to connect the scattering angle of the X-rays with the size and orientation of unit-cells in the crystalline structure.

The application to chemistry was immediately apparent. This wasn't quite an idea out of the blue. Ewald had previously thought of using diffraction to look at the structure of crystals, but realized that the wavelength of visible light is too large for the job, being greater than the space between the atoms or molecules in a crystal lattice. The shorter wavelength of X-rays was just right, as it was about the same as the spacing within the crystal.

The first crystalline structure to be elucidated by X-ray crystallography was table salt (sodium chloride) in 1914, establishing the existence of ionic compounds (see page 136). The structure of diamond was shown by William Bragg in 1913, confirming the tetrahedral bonding of carbon proposed by van 't Hoff. In the 1920s, X-ray crystallography was used to

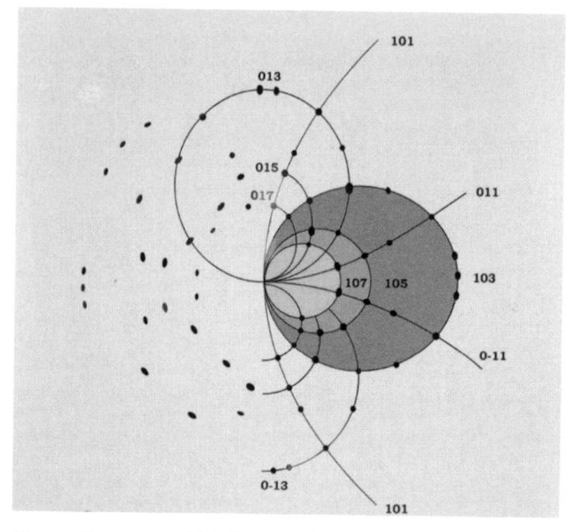

Bragg interpreted his X-ray photographs of crystalline structures by measuring distances and angles to work out the positions of atoms.

A ball-and-stick model showing the molecular structure of penicillin, discovered by Dorothy Crowfoot Hodgkin in 1946. Green, white, red, yellow and blue balls represent atoms of carbon, hydrogen, oxygen, sulphur and nitrogen respectively. The sticks represent the bonds between them.

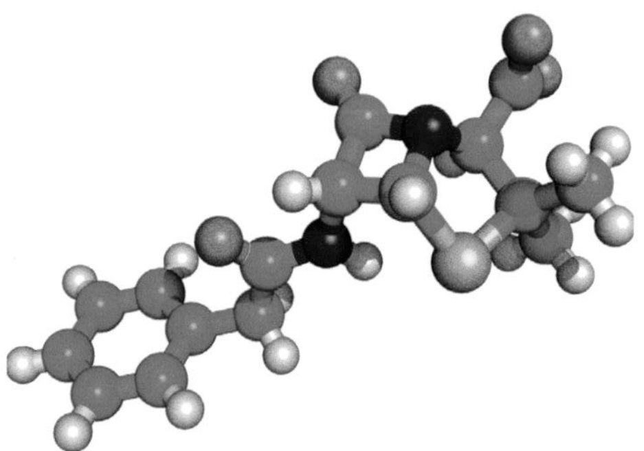

Modern computer rendering of the structure of the penicillin molecule.

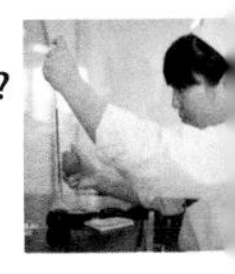

DOROTHY CROWFOOT HODGKIN (1910–94)

Born Dorothy Crowfoot in Egypt, Hodgkin spent her childhood in Norfolk and the Sudan. Her state school allowed her to study chemistry with the boys (the subject was not usually available to girls) and provided extra-curricular Latin lessons so that she could apply to the University of Oxford. (Latin was a requirement of entrance at the time.) She was only the third woman to gain a first-class degree from Oxford. She became interested in the use of X-ray crystallography to investigate biological molecules, and was involved in the first such work, on the structure of pepsin, a digestive enzyme.

Hodgkin first encountered vitamin B_{12} in 1948, the year of its discovery, and grew new crystals of it. Its structure was unknown at the time. She recognized that it contains cobalt, but nothing else could be deduced, making it the most challenging substance to tackle with X-ray crystallography. She published the final structure in 1955, and in 1964 became the first (and, so far, the only) British woman to receive a Nobel Prize in science.

Her work on insulin began in 1934 when she received a small sample of crystalline insulin. X-ray crystallography was not then sufficiently advanced to work with insulin; Hodgkin contributed to the development of the technique until, in 1969, she finally revealed its structure.

discover the arrangement of atoms within minerals and metals and their compounds. Linus Pauling found the structure of magnesium stannide (Mg_2Sn) in 1923, and garnet (a family of silicate minerals) became the first mineral structure to be explored in 1924. By showing the spaces between atoms in a matrix, X-ray crystallography also revealed the sizes of atoms and the lengths of various bonds.

Life-chemistry and X-ray crystallography

During the 1920s and 1930s, X-ray crystallography was refined and perfected and began to be applied to organic compounds. The first organic molecules were investigated with the technique in the 1920s. These were relatively small, starting with hexamethylenetetramine (a combination of formaldehyde and ammonia), studied in 1923. This was followed by some long-chain fatty acids. The first large organic molecules were investigated in the 1930s, with the most important work carried out by the British chemist Dorothy Crowfoot Hodgkin (1910–94). She worked out the structures of cholesterol (in 1937) penicillin (in 1946), vitamin B_{12} (in 1956) and insulin (in 1969) – the last of these took her more than 30 years.

The big one – DNA

The path to unravelling DNA as the chemical that carries genetic coding was a long and convoluted one, only part of which belongs to the story of chemistry. The long strands of chromosomes within living cells were first noticed but not correctly identified in the 19th century. Their constituent material (DNA with protein and RNA) was named both nuclein and chromatin at different times. Albrecht Kossel showed in 1878 that 'nuclein' contains a non-protein component that he identified as nucleic acid. Between 1885 and 1901 he identified the nucleotide bases of DNA and RNA: adenine, cytosine and guanine (found in both), thymine (found in DNA only), and uracil (found in RNA only). The Russian-American biochemist Phoebus Levene identified the base, sugar and phosphate nucleotide unit in 1919 and suggested that DNA is a string of nucleotide units linked by phosphate groups. He did not suggest variety in the sequence, though, assuming instead that the structure just repeated similar units. This would not give much potential for the molecule to carry any kind of complex coding and so made it look less likely than previously that DNA carried genetic coding.

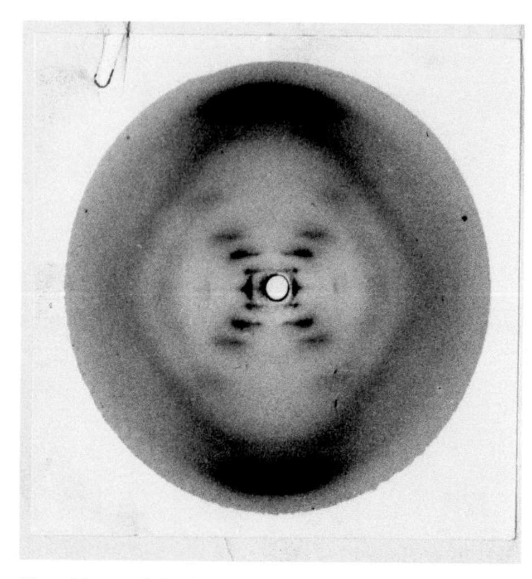

Franklin and Gosling's Photo 51, the key to unlocking the structure of DNA.

The Austrian chemist Erwin Chargaff found in the late 1940s that the bases in DNA always occur in pairs, and in 1944 Oswald Avery established that DNA carries genetic information, a finding confirmed in 1952 by Alfred Hershey and Martha Chase. But the structure of DNA itself remained elusive. Before using X-ray crystallography, chemists knew only that there were units comprising a nucleotide base attached at right-angles to a sugar and phosphate group. They knew these units could link together to form a chain, but didn't know how many were involved, or the structure that the linked chain or chains made. It was possible that DNA comprised one, two, three or more chains somehow linked together, that the phosphates grouped in the middle or stuck out at the sides. Everything was up for grabs.

The race to find the structure narrowed to a team in Cambridge, England, and the chemist Linus Pauling working in the USA. Pauling published his attempt in 1953, proposing a three-chain helix with a phosphate-sugar backbone in the centre. The British team, consisting of Francis Crick and James Watson, with the help of Maurice Wilkins in London, had to act quickly before others pointed out the errors in Pauling's structure that were clear to them – his model would not act as an acid, so could not be correct. Wilkins gave Crick and Watson the crucial piece of the puzzle in a distinctly unethical act. The vital element was an excellent X-ray photograph ('Photo 51') taken in 1952 by PhD student Raymond Gosling under the supervision of X-ray crystallographer Rosalind Franklin. But Wilkins showed it

to Crick and Watson without Franklin's permission.

Photo 51 was much clearer than those they had used previously and than any photo available to Pauling. It showed the double helical nature of the DNA chain with its backbone of alternating deoxyribose and phosphate molecules. From that, Crick and Watson were able to make calculations that determined the overall size and structure of DNA: a double helix, with sugar-phosphate groups making up the sides of the ladder and paired bases making up the rungs.

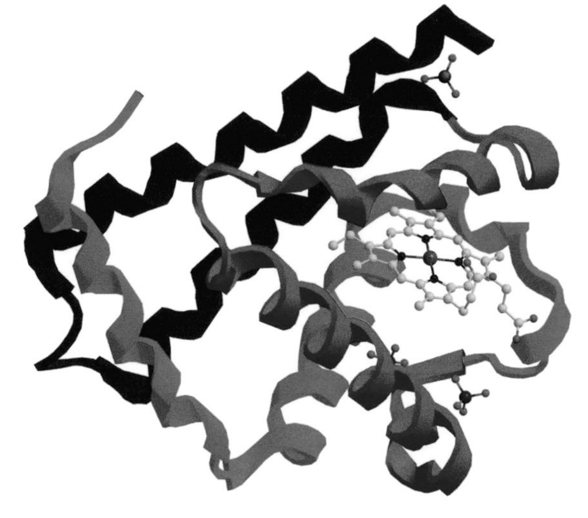

Ribbon model of myoglobin, a protein which binds oxygen in muscle. This type of model focuses on the shape rather than the chemical composition of the molecule

Discovering the structure of DNA opened the door for modern genetics, the mapping of the human genome, genetic engineering and genetic medicine. It also proved beyond any shadow of a doubt that everything in life resolves to chemistry.

Proteins unravelled

In the second half of the 1950s, chemists began to work out the structures of proteins using X-ray crystallography. Proteins are large and complex molecules that are essential to all life processes. The first protein structure to be discovered was myoglobin from a sperm whale; since then, more than 86,000 macromolecular structures have been determined using X-ray crystallography – ten times as many as with the next most popular method.

Proteins have an irregular shape that determines the protein's behaviour and is fundamental to its function in life systems. When proteins change shape, they 'denature' and their character and capabilities change. A common example of denatured protein is cooked egg white – the changed protein can't be reformed to its original configuration (you can't uncook an egg). Today, understanding the shape and functions of proteins is giving insight into areas as diverse as nutrition and the action of viruses.

HOW TO DO IT

Modern X-ray crystallography works best with a single, very pure crystal. This is slowly rotated while the pattern of X-ray diffraction is recorded from different angles. Many datasets are gathered. A computer is then used to process the images and work out the bond lengths, bond angles and locations of different atoms in the crystal and build a functional three-dimensional model of the molecule.

CHAPTER 9

MAKING STUFF

'He who knows forms grasps the unity of nature beneath the surface of materials which are very unlike. Thus is he able to identify and bring about things that have never been done before, things of the kind which neither the vicissitudes of nature, nor hard experimenting, nor pure accident could ever have actualised, or human thought dreamed of.'

Francis Bacon, *Novum organum*, aphorism 3, 1620

No longer the arcane province of alchemy, the transformation of matter lies at the heart of vast chemical industries. We not only manufacture chemicals we need, we can even design in the properties we want, fine-tuning molecular structures to create substances that don't occur in nature at all.

Neoprene, invented in 1930, is a synthetic rubber with a wide range of uses from sub-aqua gear to insulation.

Synthesis and synthetics

People have been making things with chemistry since pre-history. Perfumes, glazes, even soup, are all products of people working with chemistry. We combine and process chemicals in ways that could never happen in nature to produce new materials. Many have come about accidentally or through trial and error, but then found a use and sometimes spawned whole groups of further new materials. One of the most prolific and important groups is plastics.

Plastics revolution

In common usage 'plastic' is the type of material we associate with drinks bottles, food trays, lunch boxes and toys – a hard or bendy material, often brightly coloured, that softens and burns on heating. To a chemist, though, a plastic is an organic material that is pliable when soft but retains its shape when it hardens. There are naturally occurring plastics, such as amber and rubber, but the artificial plastics are now more widely known and used. All plastics are polymers: their molecules are built up from repeating units into chains at least 1,000 units long. The properties of the plastics depend on their molecular structure, with single, unbranched chains tending to produce slippery and viscous substances, while cross-links between chains add strength.

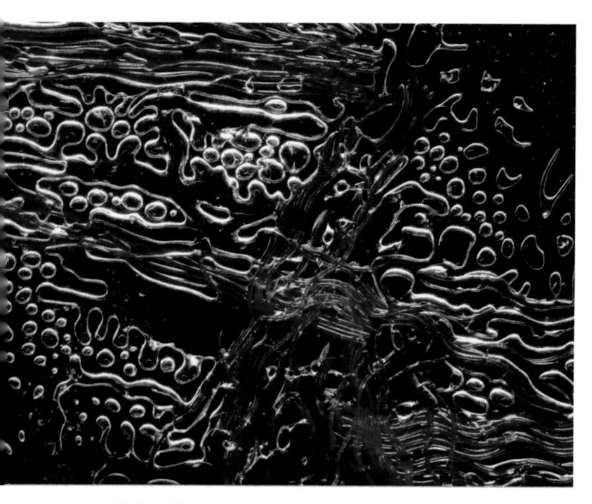

Nylon fibres seen through a microscope; the area shown in the image is 1mm across.

Starting from nature

The first plastics used were naturally occurring substances. People have been using amber and latex (the sap from the rubber tree) for millennia. During the 19th century, chemists began to modify some of the naturally occurring polymers to make them more useful. Rubber and cellulose (the material that makes up plant fibres) were both adapted. This set the stage for the development of fully artificial plastics.

Rubber to the rescue

From 1832 to 1834, Nathaniel Hayward and Friedrich Ludersdorf discovered that mixing rubber with sulphur removed the stickiness that had made it difficult to use. Hayward probably told Charles Goodyear, who in 1845 patented the vulcanization process that renders rubber durable and non-viscous. It involves adding a curative (initially, and often still, sulphur) and then heating the rubber under pressure to create cross-links between the strands. Three weeks before Goodyear's US patent, Thomas Hancock acquired a British patent for the same process. Goodyear profited little; others, however, made millions from his invention. His latter years were spent trying and mostly failing to protect his patents. He was

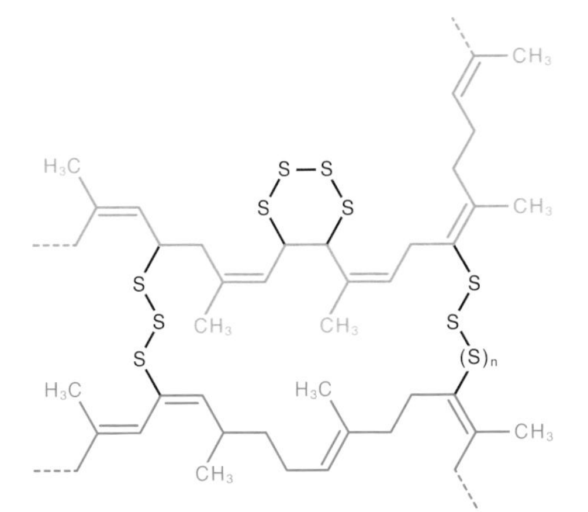

The molecular structure of rubber. Sulphur atoms form bonds between the hydrocarbon chains.

RUBBER PEOPLE

The first civilization to arise in South America was that of the Olmec, whose name means 'rubber people'. They took latex from the rubber tree and treated it with the juice from a local vine to create processed rubber as early as 1600BC.

An early rubber ball from Peru, c. *1650, with seeds from the rubber tree.*

imprisoned for debt in 1855 and died five years later owing $200,000. The Goodyear Tire and Rubber Company, founded years after his death, is named after him.

Rubber can be left with some bendiness (as in Wellington boots) or can be made hard. Its most useful – indeed, world-changing – application sounds remarkably mundane. Rubber gaskets and seals were used in the machinery that drove the Industrial Revolution. Previously, strips of leather soaked in oil had been used to plug gaps and make seals in mechanical joints and connections, but rubber was much better. It was flexible enough to be compressed between moving parts but regained its shape afterwards and could be moulded to fit exactly where it was needed. It was more durable than leather, and produced less friction in the moving machinery. In the steam-powered engines of the 19th century, and later the oil-and-petrol-driven vehicles of the 20th century, the rubber gasket became an invaluable component, greatly enhancing efficiency and making it possible for machines to carry out tasks that would have been impossible without it.

More plastics from plants

As Goodyear was experimenting with latex from the rubber tree, two French chemists, Louis-Nicolas Ménard and Florès Domonte, were working with nitrocellulose (or guncotton) – that is, cellulose which had been nitrized by exposure to nitric acid. They found it to be soluble in ether and that if they added ethanol a clear, gelatinous liquid resulted that could be painted on to human skin to dry to a flexible film. It was used as a wound dressing from 1847 and

A photographer in France, 1890. Her film would have used collodion.

named collodion. A completely different use for collodion was discovered in 1851, when the English sculptor Frederick Scott Archer found that it could be used to make photographic film, eventually replacing the daguerreotype method.

Another Englishman, Alexander Parkes, noticed that a white residue remained when photographic collodion evaporated. With that discovery, he started the plastics industry. He began to manufacture the substance under the name Parkesine, selling it as a waterproofing agent for cloth.

John Hyatt formed the Albany Billiard Ball Company in 1868 to make billiard balls from celluloid.

Unfortunately, his business went bankrupt while trying to upsize to meet demand.

The next development took place in the USA and UK at around the same time, leading to a legal battle over priority and patents. Daniel Spill in the UK and John and Isaiah Hyatt in the USA both added camphor to cellulose nitrate, making a product that Spill named xylonite and Hyatt called celluloid. The second name is still in use. It was a hard plastic that resembled ivory or horn and was used as a cheap replacement for those substances as well as for many other hard objects.

Celluloid was used for film until the 1950s when it was replaced by acetate. In keeping with its origins in guncotton, celluloid is highly flammable and self-ignites at temperatures above 150° C (easily reached in front of a projector). Acetate is much safer.

New plastics

We tend to think of plastics as a 20th-century development, but both PVC and polystyrene were first made accidentally in the first half of the 19th century. Neither was recognized as useful at the time. In 1835, the French chemist and physicist Henri Regnault left a sample of vinyl chloride gas sitting in sunlight. He later found a white solid at the bottom of the flask. The vinyl chloride had formed a polymer – polyvinyl chloride, or PVC – the molecules joining together into a long chain.

Regnault could see no use for PVC and it languished unexploited for the best part of a century before the American chemist Waldo Semon found out how to plasticize it (that is, make it bendier) with additives.

The more flexible material was immediately used for shower curtains and soon a whole host of other products.

Polystyrene got off to a similarly unpromising (and accidental start). It was discovered in 1839 when the German pharmacist Eduard Simon was trying to distil a natural resin, called storax, and obtained an oily substance that he called 'styrol'. Over the next few days, his oil thickened, naturally forming polystyrene. It took until 1920 and another German chemist, Hermann Staudinger, to realize that polystyrene is simply a chain of styrene molecules. Commercial manufacture of polystyrene began in 1930.

But these early discoveries were overshadowed by the first artificial plastic to be recognized as useful: Bakelite. Setting out to make a substitute for shellac, a natural resin-like substance, Belgian chemist Leo Baekeland investigated reactions between phenol and formaldehyde, inadvertently making the first true plastic. The discovery of Bakelite was formally announced in 1909; and it was transformative. Bakelite doesn't melt, distort or discolour and is an electrical and thermal insulator. It quickly found many uses. When it began to replace the wood in wireless sets in the 1930s, radios suddenly became an affordable item for the general public. Bakelite also sparked a revolution in plastics: suddenly they looked interesting, useful and new.

Bakelite radios kickstarted the mass communications revolution of the 20th century.

The 1930s proved to be a turning point in polymer development. Polyethylene followed polystyrene in 1935, and in 1937 Wallace Carothers invented nylon, intended initially as an alternative to silk. When nylon stockings were launched in New York in 1940, four million pairs were sold in just a few hours.

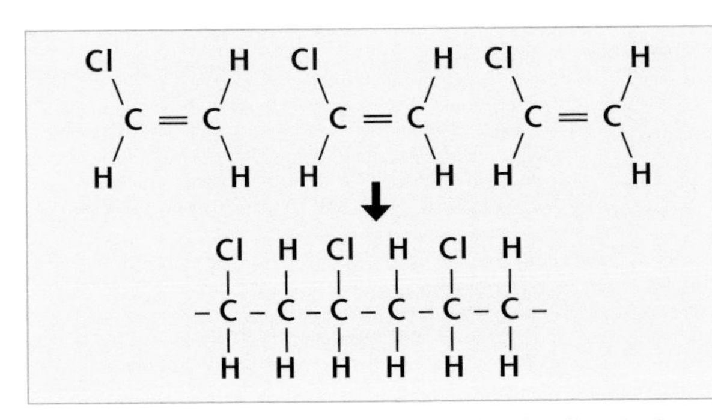

PVC is made by breaking the double carbon bond within vinyl chloride molecules (C_2H_3Cl) and stringing the groups together into a long chain.

Designer plastics

Polyethylene and polypropylene (polyolefins) account for almost half the plastics sold in the USA each year. Since the 1950s, metal catalysts including titanium and vanadium have been used to break the double bonds between carbon atoms in ethylene and propylene and prompt the formation of chains (polymerization). But

ALL PUFFED UP

Expanded polystyrene, familiar now as packing and insulation material, was developed in 1941 and consists of 98 per cent air. One of its first uses was to make six-person life-rafts for the US Coast Guard in 1942. It has since become a considerable environmental pollutant as it is widely used and difficult to dispose of. In 2017, the US state of California introduced a wide-ranging ban on the use of expanded polystyrene in disposable products.

their action is not precise and it's difficult to produce pure plastics. In the 1990s, chemists began experimenting with new types of catalyst to try to gain better control over the design and production of plastics. These include new organo-metal catalysts and metallocenes.

Discovered in 1953, the metallocene molecule has two halves made of a ring of five carbon atoms, which enclose a positively charged metal ion. As single-site catalysts, they offer precise control over polymerization and their bonds can be fine-tuned to create the properties needed. For example, packaging film for food can be tailored to have different porosities allowing different foods to 'breathe'. Metallocenes can also be used to combine monomers that are normally considered incompatible.

War and necessity, mothers of invention

While some important materials and processes were discovered by accident, others were the result of a dedicated search prompted by need. In particular, the wars of the 20th century disrupted the supply of many essential chemicals to Europe, prompting a search for artificial alternatives.

During the Second World War, the manufacture of parachutes from natural and, subsequently, artificial fibres was of vital importance.

Keep on trucking

Artificial rubber was first created in the form of polymerized butadiene in 1910 by the Russian chemist Sergei Lebedev. With supplies of rubber threatened during the First World War, there was a race to increase its production. This was not achieved until 1928, too late for the First World War but in good time for the Second. The raw material was ethanol, distilled from grain or potatoes. By the start of the Second World War, major powers including Japan, Germany, the Soviet Union and the USA all had factories producing synthetic rubbers that could be used for tyres. German factories became the target of Allied bombing attacks in an attempt to cripple vehicle and aircraft production.

Similarly, nylon and other artificial fibres were produced to replace threatened silk supplies. Although first used for stockings and underwear, production of nylon was rapidly channeled into making items such as parachutes and ropes needed for the Second World War.

Plants and explosives

Farmers have known for a long time that human and animal waste and carcasses are beneficial for the soil and help crops to grow. In many parts of the world, night soil (human waste) was spread on fields, and horse manure was also widely used. The German chemist Justus von Liebig (see page 143), who did extensive work on plant nutrition, accused Britain of stealing 3.5 million skeletons from Europe to grind up as bone meal to feed crops. By the late 19th century it was clear that the vital ingredient in fertilizer is nitrogenous compounds.

From 1820 to around 1860, Peru exported guano (bat and bird droppings) from the Chincha islands to Europe and America, where it served both as fertilizer and as the source of the nitrogen compounds used to make explosives. When the 12.5 million tonnes of guano was exhausted, another source had to be found. It turned out to be the saltpetre deposits of the Atacama Desert and a war ensued over control of the area, won eventually by Chile. Over the coming years, Chile grew rich on saltpetre exports and by 1900 produced two thirds of the fertilizer in use around the world. It was clear that this supply, too, would eventually run out.

The problem seemed particularly pressing to German chemists. Germany had poor, infertile soil and imported a great deal of Chilean saltpetre. Food security was threatened, but that was not all. The German chemist and Nobel laureate Wilhelm Ostwald pointed out that the nitrogen shortage was also a threat to national security as it was needed for the manufacture of explosives. In the arms race that led up to the First World War, this threat was compelling.

It was already known that ammonia is a compound of nitrogen, and that nitrogen is present in the atmosphere, but attempts to fix nitrogen from the air had failed. The problem was solved by Fritz Haber. In 1905, while experimenting with the thermodynamics of gases, he passed nitrogen and hydrogen over an iron catalyst at a temperature of 1000°C and produced a small quantity of ammonia in the process. By 1909, he had perfected the production of ammonia, finding that if he increased the pressure to 150–200 atmospheres he could

Making fertilizer following the Haber–Bosch process at the I.G. Farben fertilizer production plant in Germany, 1930.

reduce the temperature to 500 degrees C. The method was industrialized with the help of Carl Bosch in 1913 and is now known as the Haber–Bosch process. In enabling the continued German production of explosives, it extended the First World War. But it also provided a virtually limitless supply of fertilizer when adapted to produce ammonium sulphate. The Haber–Bosch process still produces half the fertilizer used around the world.

Atom by atom

With the advent of powerful computers and increasing knowledge about the nature of chemical bonds and the dynamics of reactions, designing new chemical products has become a highly sophisticated high-tech process. Most designer molecules are organic, including the vast output of the pharmaceutical industry. Chemists work out the content and structure of proteins and other biological molecules and then design other molecules that will lock into them, and either inhibit or enhance their function. These techniques are beyond the scope of the present book. But at the other end of the scale, innovative materials can be made even from a single element, just by controlling the positions of atoms.

Carbon in all shapes

Controlling the positions of atoms is key to exciting work in exploiting the allotropes of carbon. In addition to the naturally occurring graphite and diamond, carbon can be manipulated into forming spheres of Buckminster fullerene, graphene and nanotubes (long sheets of graphene rolled into tubes). Buckminster fullerene (C_{60}) was first observed in 1985 by Richard Smalley, Bob Curl and Harry Kroto in a joint US/UK research project that involved vaporizing

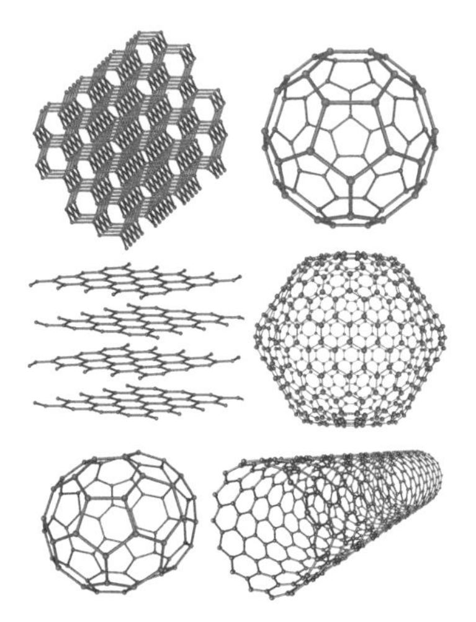

Allotropes of carbon, clockwise from top right: Buckminster fullerene C_{60}; C_{70}; graphene nanotube; C_{20}; graphene sheet; diamond.

carbon with a laser beam and using mass spectrometry (see page 186) to investigate the resulting plasma. They found that a molecule comprising 60 carbon atoms was always among the products and deduced that the structure is a hollow sphere, informally known as a 'buckyball'. The surface is composed of hexagonal and pentagonal faces. C_{70} and C_{20} versions also form, though C_{60} dominates.

Graphene has been called the first two-dimensional material, existing as a sheet of carbon a single atom thick. The atoms are arranged in a hexagonal lattice, as they are in graphite – graphene is a single layer of graphite. It was first identified by Andre Geim and Kostya Novoselov working in Manchester, England, though it had been discussed since 1947. Geim and Novoselov had informal lab sessions on Friday evenings at which they tried out experiments not linked to their regular work. On one occasion, they discovered that they could use sticky tape to remove very thin layers from a block of graphite. By repeatedly thinning the flakes of graphite they eventually made a sheet only one atom thick.

Graphene has extraordinary properties: it's stronger than any other material, very light, very flexible and is the best conductor of heat and electricity known. It can be rolled into tubes – nanotubes – of varying diameter but with walls a single atom thick. Nanotubes have been made with a length:diameter ratio of 132,000,000:1.

Carbon revolution

The uses to which these allotropes of carbon can be put are just beginning to be explored. Buckminster fullerene can be used as a cage, with another molecule trapped inside, and has potential as a delivery system. It could be used to carry drugs directly to a targeted area of the body, such as a tumour. Nanotubes form a super-strong structure that can be used to reinforce other materials, but could potentially make super-thin, super-light, semi-transparent electronics. Graphene membranes could be made into filters to provide clean water or in desalination technology to harvest drinking water from the sea.

SECRET STRENGTH

There are many tales of incredibly sharp swords made of Damascus steel. The blades have a distinctive swirly pattern that is a feature of the wootz steel from which they were made in the Middle East. Wootz steel came from India, where it was first made around 600BC. It has not been produced since 1750 and the production methods used have been lost; attempts to recreate it have failed.

German materials scientist Peter Paufler suggested in 2006 that carbon nanotubes formed naturally in wootz steel from the plant material used in forging the steel, and these gave the Damascus blades their extraordinary properties. The nanotubes would have filled with cementite, an iron-carbon compound.

INTO THE FUTURE

Today, chemistry is intricately connected with many other sciences, including biology, pharmacology, medicine, materials science, physics, geology and astronomy. Its core concerns – identifying, understanding and synthesizing chemicals – are universally relevant. Over the coming decades, we can expect to see great strides in the application of nanostructures, engineered molecules, genetic engineering, energy production and the synthesis of pharmaceuticals and foodstuffs. And there will quite likely be advances we can't begin to foresee.

Chemical solutions to chemical problems

Chemistry has brought huge benefits to humankind, but in their wake have come immense problems. Among them are pollution, global warming, drug-resistant

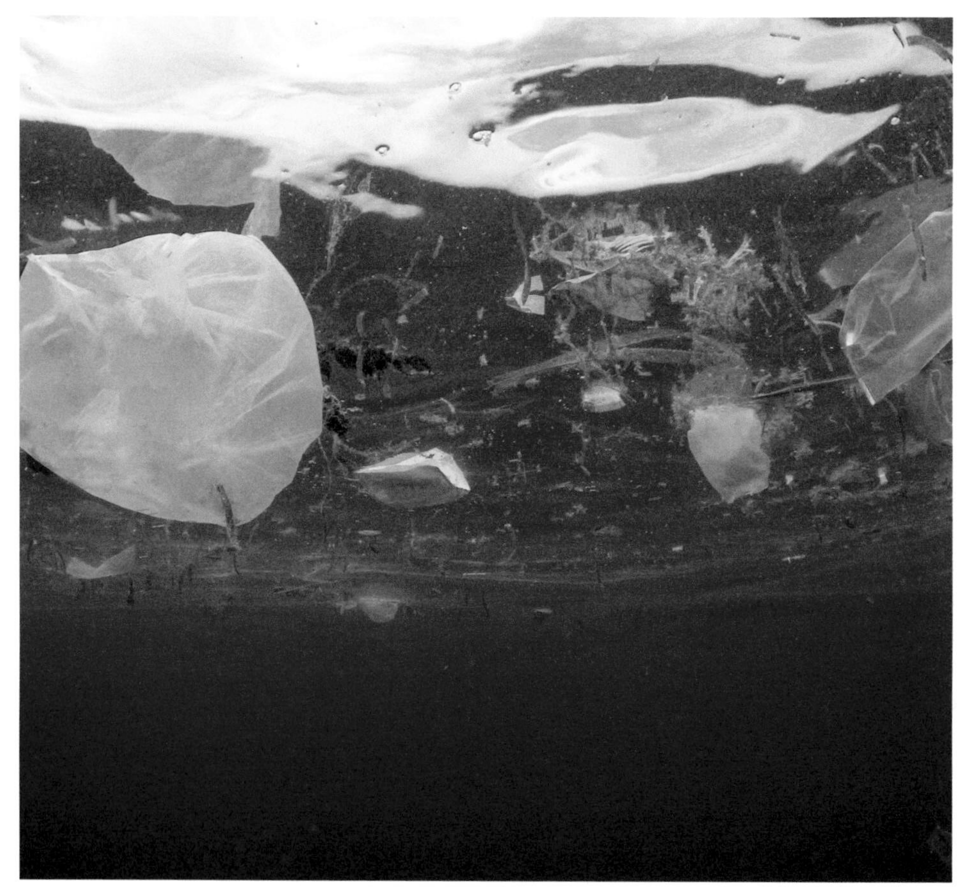

Plastic bags and other rubbish accumulate in the sea, endangering wildlife and threatening ecological balance.

Microorganisms in mealworms' guts break down expanded polystyrene during digestion. Scientists found that the worms converted about half of the carbon in the polystyrene they had eaten into carbon dioxide, as they would with other food sources. The other half was excreted.

microbes and decreasing biodiversity. We can't turn the clock back, but we can begin to use science to solve some of the problems its misuse has brought about.

The use of fossil fuels has increased exponentially since the start of the 20th century. Coal, oil, gas and their products are used extensively for energy and for the manufacture of plastics and other hydrocarbon-based materials with the result that we are in danger of running out of them. While burning fossil fuels adds extra carbon dioxide to the atmosphere, contributing to climate change, locking fossil-carbon up in plastics that can take millennia to biodegrade brings other environmental challenges.

The quest for cleaner energy can be addressed by chemists and physicists. One possible chemical solution is the use of hydrogen fuel cells, which produce energy by combining hydrogen and oxygen, generating only water as a by-product. Some forms of pollution could be tackled by enzymes originally produced by living organisms but that might be manufactured en masse. In 2015, researchers at the University of Beijing, China, found bacteria in the guts of mealworms that could digest expanded polystyrene. And in 2016, researchers at the University of Kyoto in Japan found a bacterium, *Ideonella sakaiensis*, that will feast on PET, a form of plastic widely used for food containers. The researchers isolated the enzyme it produces and made more of it, which they used successfully to break down PET in the laboratory.

Learning from the past

The great advances in civilization have all exploited chemical knowledge. With the benefit of hindsight, it's possible to see where we should have used caution in some of our endeavours with chemistry. More advances doubtless lie ahead of us; let's hope we use them wisely.

INDEX

PICTURE CREDITS

Front cover: bottom left and bottom right, Wellcome Library, London; centre, Science Photo Library/Medi–mation; all other images, Shutterstock

Alamy Stock Photo: 13 bottom (Ancient Art and Architecture), 92, 132 left (Universal Images Group North America LLC), 171 bottom (World History Archive)
Bridgeman Images: 30, 33, 34 top, 43, 46, 49, 52, 57, 70, 80, 82, 85, 142 bottom
Diomedia: 50 (Science Source/New York Public Library Picture Collection)
Getty Images: 8, 60 (SuperStock RM), 72 (DeAgostini), 77, 99, 128, 134 (Bettmann Archive), 139 (The LIFE Picture Collection), 153 (ullstein bild), 164 (AFP), 167 (The Print Collector/Heritage–Images), 168–9 (Bloomberg), 180 (Corbis), 196 top (Corbis), 198 (Popperfoto Creative), 200 top (ullstein bild)
Mary Evans Picture Library: 10–11 (Interfoto/Sammlung Rauch), 24–5 (Interfoto/Bildarchiv Hansmann), 54–5 (Photo Researchers), 68–9 (Photo Researchers), 76 bottom (Photo Researchers)
NASA: 101, 135
Sandbh: 130
Science & Society Picture Library/Science Museum, London: 17 top, 31, 41, 45, 113, 126, 188 bottom left, 195 bottom
Science Photo Library: 89 (Sheila Terry), 121, 125 (Charles D. Winters), 157 (Sheila Terry), 184 top, 188 top (Ramon Andrade 3Dciencia), 189
Shutterstock: 6, 7, 12, 13 top and centre, 17 bottom, 18 bottom, 20, 22, 27 top, 37x2, 44 bottom, 63, 76 top, 79, 81, 86–7, 88x2, 93 bottom, 99 top, 108, 109, 110, 118, 119, 122, 132 right, 140–141, 142 top, 145, 151x2, 154, 160 bottom, 163, 165, 166, 174, 176, 177, 178, 179, 182, 184 bottom, 191, 192–3, 196 bottom, 197 top, 201, 202, 203
US National Library of Medicine: 64
University of Oregon: 190
Wellcome Library, London: 47, 59, 62, 65, 78, 91, 93 top, 94, 95, 96, 98, 103, 104x2, 111, 116, 120, 124, 133, 144, 146, 148 right, 155, 158, 159, 161, 170, 171 top, 172, 173 top, 175, 181, 185, 186, 194 (Macroscopic Solutions)
Wikimedia/Swetapadma07: 19

Artworks on pages 44 top, 152 top right, 187 top and bottom left, and 195 top, by David Woodroffe.